T0135795

Mathematical Physics III. Integrable Systems of Classical Mechanics. Lecture Notes

Matteo Petrera

Institut für Mathematik, MA 7-2
Technische Universität Berlin
Strasse des 17. Juni 136

Bibliographic information published by the Deutsche Nationalbibliothek

The Deutsche Nationalbibliothek lists this publication in the Deutsche Nationalbibliografie; detailed bibliographic data are available in the Internet at http://dnb.d-nb.de .

ISBN 978-3-8325-3950-4

Logos Verlag Berlin GmbH
Comeniushof, Gubener Str. 47,
10243 Berlin
Tel.: +49 (0)30 42 85 10 90
Fax: +49 (0)30 42 85 10 92
INTERNET: http://www.logos-verlag.de

Motivations

These Lecture Notes are based on a one-semester course taught in the Winter semester 2014/2015 at the Technical University of Berlin, to bachelor/master undergraduate mathematics and physics students.

These Lecture Notes are based on the references listed on the next pages. It is worthwhile to warn the reader that there are several excellent and exhaustive books and monographs on the topics that were covered in this course. A practical drawback of some of these books is that they are not really suited for a 4-hours per week, one-semester course. On the contrary, these notes contain only those topics that were actually explained in class. They are a kind of one-to-one copy of blackboard lectures. Some topics, some aspects of the theory and some proofs were left out because of time constraints.

A characteristic feature of these notes is that they present subjects in a synthetic and schematic way, thus following exactly the same pedagogical strategy used in class. Notions, concepts, statements and proofs are intentionally written and organized in a way that I found well suited for a systematic and effective understanding/learning process.

The aim is to provide students with practical tools that allow them to prepare themselves for their exams and not to substitute the role of an exhaustive book. This purpose has, of course, drawbacks and benefits at the same time. As a matter of fact, many students wish to have a "product" which is readable, compact and self-contained. In other words, something that is necessary and sufficient to get a good mark with a reasonable effort. This is - at least ideally - the positive side of good Lecture Notes. On the other hand, the risk is that their understanding might not be fluid and therefore too confined. Indeed, I always encourage my students to also consult more "standard" books like the ones quoted on the next pages.

Aknowledgments

The DFG (Deutsche Forschungsgemeinschaft) collaborative Research Center TRR 109 "Discretization in Geometry and Dynamics" is acknowledged. I am grateful to Yuri Suris and to the BMS (Berlin Mathematical School) for having given me the opportunity to teach this course. Special thanks go to Jan Techter (TU Berlin) for his careful proof-reading of the text.

Matteo Petrera
Institut für Mathematik, MA 7-2, Technische Universität Berlin
Strasse des 17. Juni 136
petrera@math.tu-berlin.de
May 21, 2015

Books and references used during the preparation of these Lecture Notes

Ch1 Manifolds, Lie groups, Lie algebras

- ✓ [AdMoVa] M. Adler, P. van Moerbeke, P. Vanhaecke, *Algebraic Integrability, Painlevé Geometry and Lie Algebras*, Springer, 2004.
- ✓ [LaPiVa] C. Laurent-Gengoux, A. Pichereau, P. Vanhaecke, *Poisson Structures*, Springer, 2013.
- ✓ [MaRa] J. Marsden, T. Ratiu, *Introduction to Mechanics and Symmetry*, Springer-Verlag, 1999.
- ✓ [Ol] P.J. Olver, *Applications of Lie Groups to Differential Equations*, Springer, 1998.
- ✓ [Pet] M. Petrera, *Mathematical Physics I. Dynamical Systems and Classical Mechanics. Lecture Notes*, Logos Verlag, Berlin, 2013.

Ch2 Poisson structures

- ✓ [AdMoVa] M. Adler, P. van Moerbeke, P. Vanhaecke, *Algebraic Integrability, Painlevé Geometry and Lie Algebras*, Springer, 2004.
- ✓ [LaPiVa] C. Laurent-Gengoux, A. Pichereau, P. Vanhaecke, *Poisson Structures*, Springer, 2013.
- ✓ [Pet] M. Petrera, *Mathematical Physics I. Dynamical Systems and Classical Mechanics. Lecture Notes*, Logos Verlag, Berlin, 2013.
- ✓ [Se] M.A. Semenov-Tian-Shansky, *Integrable Systems: the r-matrix approach*, Lecture Notes available at `http://www.kurims.kyoto-u.ac.jp/preprint/file/RIMS1650.pdf`.
- ✓ [Su] Yu.B. Suris, *The Problem of Integrable Discretization: Hamiltonian Approach*, Birkhäuser Verlag, 2003.

Ch3 Completely Integrable Systems

- ✓ [AbMa] R. Abraham, J. Marsden, *Foundations of Mechanics*, Addison-Wesley Publishing Company, Inc., 1987.
- ✓ [AdMoVa] M. Adler, P. van Moerbeke, P. Vanhaecke, *Algebraic Integrability, Painlevé Geometry and Lie Algebras*, Springer, 2004.
- ✓ [LaPiVa] C. Laurent-Gengoux, A. Pichereau, P. Vanhaecke, *Poisson Structures*, Springer, 2013.

Ch4 R-brackets and r-brackets

- ✓ [AdMoVa] M. Adler, P. van Moerbeke, P. Vanhaecke, *Algebraic Integrability, Painlevé Geometry and Lie Algebras*, Springer, 2004.
- ✓ [BlSz] M. Blaszak, B.M. Szablikowski, *Classical R-matrix theory for bi-Hamiltonian field systems*, Jour. Phys. A: Math. Theor. 42, 404002, 2009.
- ✓ [KoSc] Y. Kosmann-Schwarzbach, *Lie Bialgebras, Poisson Lie Groups and Dressing Transformations*, in: Integrability of Nonlinear Systems, Springer, 2004.
- ✓ [LaPiVa] C. Laurent-Gengoux, A. Pichereau, P. Vanhaecke, *Poisson Structures*, Springer, 2013.
- ✓ [Se] M.A. Semenov-Tian-Shansky, *Integrable Systems: the r-matrix approach*, Lecture Notes available at `http://www.kurims.kyoto-u.ac.jp/preprint/file/RIMS1650.pdf`.
- ✓ [Su] Yu.B. Suris, *The Problem of Integrable Discretization: Hamiltonian Approach*, Birkhäuser Verlag, 2003.

Ch5 Examples: Toda, Garnier, Gaudin, Lagrange

- ✓ [PeSu] M. Petrera, Yu.B. Suris, *An integrable discretization of the rational* $\mathfrak{su}(2)$ *Gaudin model and related systems*, Comm. Math. Phys. 283/1, 2008.

- ✓ [Su] Yu.B. Suris, *The Problem of Integrable Discretization: Hamiltonian Approach*, Birkhäuser Verlag, 2003.

Ch6 The Problem of Integrable Discretization

- ✓ [CeMcLaOwQu] E. Celledoni, R.I. McLachlan, B. Owren, G.R.W. Quispel, *Geometric properties of Kahan's method*, Jour. Phys. A: Math. Theor. 46, 2013.
- ✓ [PeSu1] M. Petrera, Yu.B. Suris, *An integrable discretization of the rational* $\mathfrak{su}(2)$ *Gaudin model and related systems*, Comm. Math. Phys. 283/1, 2008.
- ✓ [PeSu2] M. Petrera, Yu.B. Suris, *On the Hamiltonian structure of Hirota-Kimura discretization of the Euler top*, Math. Nachr. 283/11, 2010.
- ✓ [PePfSu] M. Petrera, A. Pfadler, Yu.B. Suris, *On integrability of Hirota-Kimura type discretizations*, Reg. Chaot. Dyn. 16/3-4, 2011.
- ✓ [Su] Yu.B. Suris, *The Problem of Integrable Discretization: Hamiltonian Approach*, Birkhäuser Verlag, 2003.

Siméon-Denis Poisson (1781-1840) and Joseph Liouville (1809-1882)

Marius Sophus Lie (1842-1899)

Vladimir I. Arnold (1937-2010) and Peter David Lax (1926-)

"A system of differential equations is only more or less integrable", Jules Henri Poincaré (1854-1912).

"When, however, one attempts to formulate a precise definition of integrability, many possibilities appear, each with a certain intrinsic theoretic interest", George David Birkhoff (1884-1944).

Contents

1

Manifolds, Lie groups, Lie algebras

1.1 Introduction

▶ Manifolds are, roughly speaking, abstract surfaces that locally look like linear spaces. More formally a smooth real manifold is a Hausdorff topological space that is locally diffeomorphic to the Euclidean space $\mathbb{R}^n$. Therefore it is the natural generalization of the Euclidean space $\mathbb{R}^n$, which is flat and admits a global system of coordinates.

▶ There are several reasons to generalize the Lagrangian and Hamiltonian formalism of classical mechanics defined on linear spaces to smooth manifolds. We will restrict our generalization only to the Hamiltonian formulation of mechanics, even though the Lagrangian side can be generalized as well.

▶ As a matter of fact the canonical Hamiltonian phase space $\mathbb{R}^{2n}$ (or a subset of it) has two evident limitations:

1. The *configuration (vector) space*, where coordinates $q := (q_1, \ldots, q_n)$ live, is $\mathbb{R}^n$. Many mechanical systems have a configuration space that can be described in terms of Euclidean coordinates q only at a local level. For instance, the position of a particle moving on a smooth two-dimensional closed surface embedded in the ambient space $\mathbb{R}^3$ admits a natural parametrization in terms of the coordinates that parametrize the surface. In other words, in this case, the configuration space is a surface, which is an example of smooth manifold. To fix ideas: if the surface is the (unit) sphere $\mathbb{S}^2$ the position of the particle is completely determined in terms of two angles $(\theta, \phi) \in [0, \pi) \times [0, 2\pi)$.

2. The canonical Hamiltonian phase space is by construction even-dimensional. In other words, a canonical Hamiltonian system is always described in terms of an initial value problem consisting of an even number of ordinary differential equations (ODEs). But there are systems whose description leads naturally to an odd-dimensional phase space. In other words, the mechanics is described by an initial value problem consisting of an odd number of ODEs. A natural question is to understand if such systems admit some Hamiltonian formulation.

▶ Indeed we already know some Hamiltonian systems whose configuration space is not $\mathbb{R}^n$ but a *submanifold of* $\mathbb{R}^n$. In fact, the canonical Hamiltonian formulation of mechanics is the correct local formulation of Hamiltonian mechanics when the phase space is a *symplectic manifold* (that is, maybe not surprisingly, always even-dimensional). The standard way to obtain a symplectic manifold as phase space is to start with a n-dimensional smooth manifold as configuration space. Local coordinates are $(q_1, \ldots, q_n)$. Then the $2n$-dimensional *cotangent bundle* of the manifold, locally coordinatized by the *generalized canonical coordinates* $(q_1, \ldots, q_n, p_1, \ldots, p_n)$, can be locally equipped with a symplectic structure induced by the canonical symplectic 2-form (see Example 2.10):

$$\omega_{\mathrm{can}} := \sum_{i=1}^{n} \mathrm{d}q_i \wedge \mathrm{d}p_i. \tag{1.1}$$

Example 1.1 (***Planar and spherical penduli***)

1. The planar pendulum is a Hamiltonian system with Hamiltonian

$$H(q,p) := \frac{p^2}{2} - \cos q.$$

The canonical Hamilton equations are

$$\begin{cases} \dfrac{\mathrm{d}q}{\mathrm{d}t} = \dfrac{\partial H}{\partial p} = p, \\[2mm] \dfrac{\mathrm{d}p}{\mathrm{d}t} = -\dfrac{\partial H}{\partial q} = -\sin q. \end{cases}$$

The configuration space is diffeomorphic to the circle S^1, which is a one-dimensional closed curve embedded in $\mathbb{R}^2$. Therefore, even if the ambient space of the system is two-dimensional, say the plane $\mathbb{R}^2$, the system has only one *degree of freedom*. In other words, the configuration is completely described in terms of the angle of rotation $q \in [0, 2\pi)$.

2. A *spherical pendulum* is a mass attached to a fixed centre by a rigid rod, free to swing in any direction in $\mathbb{R}^3$. The state of the pendulum is entirely determined by the position of the mass, which is constrained to move on the surface of a sphere. The configuration space is diffeomorphic to the sphere S^2. The number of degrees of freedom of the system is two.

► From Example 1.1 we argue that a possible manifestation of the first limitation is when we consider systems where the spatial configurations are subject to some geometric constraints, as for instance a rigid rod for a pendulum. One of the effects of these constraints is that the dimension of the configuration space, the so called *number of degrees of freedom*, is less than the dimension of the ambient space. Let us illustrate the meaning of such constraints.

- Consider a mechanical system in $\mathbb{R}^3$ consisting of N points. If all configurations are possible, the system is *free*. It can be described by a global system of coordinates in $\mathbb{R}^{3N}$. Then one can define a canonical Hamiltonian phase space as a subset of $\mathbb{R}^{3N} \times \mathbb{R}^{3N} \simeq \mathbb{R}^{6N}$. The dimension of the ambient space, $3N$, coincides with the dimension of the configuration space.

- However it happens quite often that there are some geometric *constraints* on the allowed configurations of the system. If such constraints do not depend on time one says that the system admits *holonomic constraints*. For example, if $N = 1$, we can require that the point lies on a given regular surface $F(x_1, x_2, x_3) = 0$

in $\mathbb{R}^3$. In such a case it is possible to introduce a local parametrization of the surface, $x_i = x_i(q_1, q_2)$, $i = 1, 2, 3$, with the property that the Jacobian matrix has maximum rank

$$\operatorname{rank}\begin{pmatrix} \frac{\partial x_1}{\partial q_1} & \frac{\partial x_1}{\partial q_2} \\ \frac{\partial x_2}{\partial q_1} & \frac{\partial x_2}{\partial q_2} \\ \frac{\partial x_3}{\partial q_1} & \frac{\partial x_3}{\partial q_2} \end{pmatrix} = 2,$$

where (q_1, q_2) vary in a certain open subset of $\mathbb{R}^2$. The vectors $\partial x / \partial q_1$ and $\partial x / \partial q_2$ are then linearly independent and form a basis in the *tangent plane* to the surface at x. Recall that a vector $\mathsf{v} \in \mathbb{R}^3$ is tangent to a surface $\mathcal{S} := \{x \in \mathbb{R}^3 : F(x_1, x_2, x_3) = 0\}$ at $p \in \mathcal{S}$ if there exists a differentiable curve $\gamma : (-\varepsilon, \varepsilon) \to \mathcal{S}$ such that $\gamma(0) = p$ and $(\mathrm{d}\gamma/\mathrm{d}t)(0) = \mathsf{v}$.

► Concerning the second limitation, a standard example of a mechanical system governed by an odd number of first-order ODEs is the three-dimensional rigid body with a fixed point (the so called *Euler top*). Let us anticipate that the Euler top is also the prototypical example of a (finite-dimensional) completely integrable system. An elementary mechanical analysis leads to a description of the body by the three components of the angular momentum (relative to body coordinates, i.e., coordinates fixed in the body). These components, say $x_1 = x_1(t), x_2 = x_2(t), x_3 = x_3(t)$, evolve according to the following system of ODEs:

$$\begin{cases} \frac{\mathrm{d}x_1}{\mathrm{d}t} = (a_2 - a_3)\, x_2\, x_3, \\ \frac{\mathrm{d}x_2}{\mathrm{d}t} = (a_3 - a_1)\, x_3\, x_1, \\ \frac{\mathrm{d}x_3}{\mathrm{d}t} = (a_1 - a_2)\, x_1\, x_2, \end{cases} \tag{1.2}$$

where a_1, a_2, a_3 are real parameters related to the inertia tensor of the body. This situation prompts two fundamental questions.

1. A configuration of the body is specified by three real numbers (to specify the rotation required to rotate the body into the given configuration). A canonical Hamiltonian description of the rigid body would use six first-order ODEs (for example, three ODEs for the evolution of three angles and three ODEs for the evolution of the corresponding momenta). However many questions are easier to study using the three ODEs (1.2). So how is the description of the Euler top related to a six-dimensional Hamiltonian description?

2. Is the dynamics Hamiltonian is some sense? We anticipate that the answer is yes. Indeed, the space coordinatized by x_1, x_2, x_3 is a prototypical example of a *Poisson manifold* and the ODEs (1.2) are already in Hamiltonian form.

► We shall see that the natural framework to develop Hamiltonian mechanics is a *Poisson manifold*, that is a smooth manifold, not necessarily of even dimension, equipped with a Poisson structure. Such a Poisson structure will allow us to properly define Hamiltonian vector fields. Remarkably a Poisson bracket on a Poisson manifold may be degenerate, unlike the canonical Poisson bracket corresponding to (1.1). But, if the bracket is non-degenerate it allows one to define locally a closed and non-degenerate 2-form, called *symplectic 2-form*. This leads to the concept of *symplectic manifold*, whose dimension is necessarily even due to the non-degeneracy condition. It turns out that every symplectic manifold is locally diffeomorphic to the canonical symplectic Hamiltonian phase space $\mathbb{R}^{2n}$.

1.2 Basic facts on smooth manifolds

► In order to define Hamiltonian dynamics on a smooth manifold we first need a formal definition of this object supplemented with the definition of the main geometrical structures one can construct on it. Before starting let us note that:

- The main geometrical features of a smooth manifold will be independent of any particular coordinate system on the open subset that might be used to define them. Therefore, it becomes of great importance to free ourselves from the dependence on particular local coordinates. From this point of view, manifolds provide the natural setting for studying objects that do not depend on coordinates.
- We will not be concerned about the degree of differentiability of all objects we are going to define: everything will be *smooth*, namely C^∞. Furthermore our presentation is restricted to the real case. Nevertheless, the generalization to complex holomorphic manifolds is possible.

► A real n-dimensional topological *smooth manifold* $\mathcal{M}$ is a set of points together with a countable collection (called *atlas*) of open sets $\mathcal{U}_\alpha \subset \mathcal{M}$ (called *local coordinate charts*), and one-to-one mappings $\chi_\alpha : \mathcal{U}_\alpha \to \mathbb{R}^n$ (called *local coordinate functions*) that satisfy the following properties:

1. The coordinate charts cover $\mathcal{M}$:
$$\bigcup_\alpha \mathcal{U}_\alpha = \mathcal{M}.$$
2. For each pair of indices α, β such that $\mathcal{W} := \mathcal{U}_\alpha \cap \mathcal{U}_\beta \neq \emptyset$, the one-to-one overlap functions $\chi_\beta^{-1} \circ \chi_\alpha$ are smooth.

Refinements of the above definition such as maximality conditions and equivalence classes of charts are not considered.

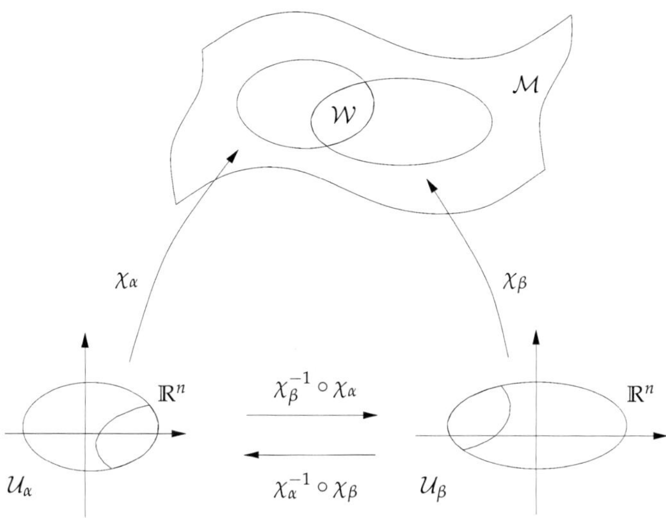

Fig. 4.1. Coordinate charts and coordinate functions.

Example 1.2 (*The Euclidean space* $\mathbb{R}^n$)

1. The simplest n-dimensional smooth manifold is the Euclidean space $\mathbb{R}^n$. There is a single ($\alpha = 1$) coordinate chart $\mathcal{U} = \mathbb{R}^n$, with a global coordinate function given by the identity map $\chi = 1_n : \mathbb{R}^n \to \mathbb{R}^n$.
2. Any open subset $\mathcal{M} \subset \mathbb{R}^n$ is a n-dimensional smooth manifold with a single coordinate chart given by $\mathcal{U} = \mathcal{M}$, with global coordinate function the identity again. Conversely, if $\mathcal{M}$ is any smooth manifold with a single global coordinate function $\chi : \mathcal{M} \to \mathbb{R}^n$, we can identify $\mathcal{M}$ with its image $\chi(\mathcal{M})$, which is an open subset of $\mathbb{R}^n$.

Example 1.3 (*The unit sphere* $\mathbb{S}^2$)

The unit sphere

$$\mathbb{S}^2 := \left\{ (x_1, x_2, x_3) \in \mathbb{R}^3 \;:\; x_1^2 + x_2^2 + x_3^2 = 1 \right\}$$

is an example of a two-dimensional smooth manifold realized as a surface embedded in $\mathbb{R}^3$.

- Let
$$\mathcal{U}_1 := \mathbb{S}^2 \setminus \{(0,0,1)\}, \qquad \mathcal{U}_2 := \mathbb{S}^2 \setminus \{(0,0,-1)\},$$
be the subsets obtained by deleting the north and south poles respectively.
- Let $\chi_\alpha : \mathcal{U}_\alpha \to \mathbb{R}^2$, $\alpha = 1,2$, be *stereographic projections* from the respective poles:
$$\chi_1(x_1, x_2, x_3) := \left(\frac{x_1}{1 - x_3}, \frac{x_2}{1 - x_3} \right), \quad \chi_2(x_1, x_2, x_3) := \left(\frac{x_1}{1 + x_3}, \frac{x_2}{1 + x_3} \right).$$
- It can be verified that the overlap function
$$\chi_1 \circ \chi_2^{-1} : \mathbb{R}^2 \setminus \{(0,0)\} \to \mathbb{R}^2 \setminus \{(0,0)\}$$
is a smooth diffeomorphism given by the inversion
$$\left(\chi_1 \circ \chi_2^{-1} \right)(x_1, x_2) = \left(\frac{x_1}{x_1^2 + x_2^2}, \frac{x_2}{x_1^2 + x_2^2} \right).$$

The unit sphere is a particular case of the general concept of a surface in $\mathbb{R}^3$.

Example 1.4 (*The torus*)

In $\mathbb{R}^3$ consider a unit circle $\mathbb{S}^1$ in the (x_1, x_3)-plane centered at $(x_1, x_3) = (1, 0)$. Then rotate it around the x_3-axis by an angle of 2π. The geometric object we obtain is a torus, an important two-dimensional smooth manifold. From the topological point of view it is the direct product of two circles $\mathbb{S}^1$. One writes $\mathbb{T}^2 \simeq \mathbb{S}^1 \times \mathbb{S}^1$. More generally, the n-dimensional torus is

$$\mathbb{T}^n \simeq \underbrace{\mathbb{S}^1 \times \cdots \times \mathbb{S}^1}_{n \text{ times}}.$$

Example 1.5 (*Implicit submanifolds of* $\mathbb{R}^n$)

Smooth manifolds often arise as zero-level sets of smooth functions:

$$\mathcal{M} := \{x \in \mathbb{R}^n \ : \ F_i(x) = 0,\, i = 1, \ldots, k\},$$

for a given set of $k < n$ smooth functions $F_i : \mathbb{R}^n \to \mathbb{R}$. If the rank of the Jacobian matrix of $(F_1, \ldots, F_k)$ is a constant ℓ with $\ell \leqslant k$ for all x, then $\mathcal{M}$ is a smooth manifold of dimension $n - \ell$. In this situation, the set $\mathcal{M}$ is called *implicit submanifold of* $\mathbb{R}^n$.

► In what follows we present the most important structures and objects that can be defined on smooth manifolds.

1.2.1 1-forms and vector fields

► Let $\mathcal{M}$ be a real n-dimensional smooth manifold and $p \in \mathcal{M}$ be a point on $\mathcal{M}$.

- The *algebra of smooth functions* on $\mathcal{M}$ will be denoted by $\mathscr{F}(\mathcal{M}) \equiv C^\infty(\mathcal{M}, \mathbb{R})$.
- The *tangent space* to $\mathcal{M}$ at p is denoted by $T_p\mathcal{M}$ and its dual space, the *cotangent space* to $\mathcal{M}$ at p, is denoted by $T_p^*\mathcal{M}$. The tangent and cotangent spaces to $\mathcal{M}$, which are n-dimensional vector spaces, form the *fibers* of the *tangent bundle* $T\mathcal{M}$, resp. the *cotangent bundle* $T^*\mathcal{M}$:

$$\begin{aligned} T\mathcal{M} &:= \bigcup_{p \in \mathcal{M}} \{p\} \times T_p\mathcal{M}, \\ T^*\mathcal{M} &:= \bigcup_{p \in \mathcal{M}} \{p\} \times T_p^*\mathcal{M}. \end{aligned}$$

  (a) The bundles $T\mathcal{M}$ and $T^*\mathcal{M}$ carry a natural structure of real smooth manifolds of dimension $2n$.

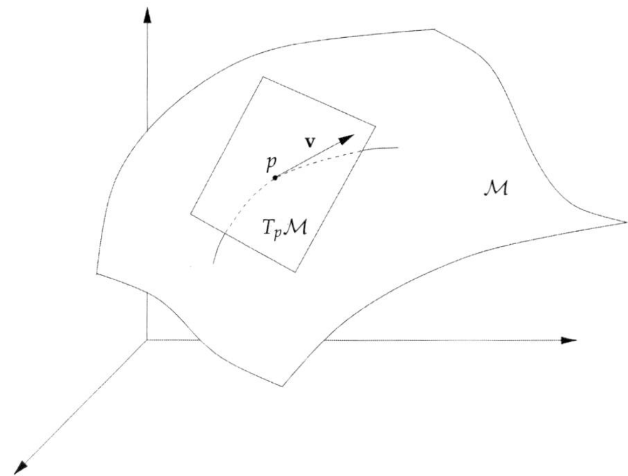

Fig. 4.2. A tangent space $T_p\mathcal{M}$.

(b) A (smooth) *vector field* v on $\mathcal{M}$ is a smooth section of the tangent bundle (i.e., a smooth map $\mathsf{v} : \mathcal{M} \to T\mathcal{M}$ that assigns a vector $\mathsf{v}(p) \in T_p\mathcal{M}$ at each point $p \in \mathcal{M}$). The $\mathscr{F}(\mathcal{M})$-module of vector fields on $\mathcal{M}$ is denoted by $\mathfrak{X}(\mathcal{M})$.

(c) A (smooth) *1-form* ω (also called *covector field*) is a smooth section of the cotangent bundle (i.e., a smooth map $\omega : \mathcal{M} \to T^*\mathcal{M}$ that assigns a 1-form $\omega(p) \in T_p^*\mathcal{M}$ at each point $p \in \mathcal{M}$). The $\mathscr{F}(\mathcal{M})$-module of 1-forms on $\mathcal{M}$ is denoted by $\Omega(\mathcal{M})$.

(d) We denote the pairing between $T_p\mathcal{M}$ and its dual $T_p^*\mathcal{M}$ by

$$\langle \cdot, \cdot \rangle : T_p^*\mathcal{M} \times T_p\mathcal{M} \to \mathscr{F}(\mathcal{M}).$$

Thus if $\mathsf{v}(p) \in T_p\mathcal{M}$ we may define a function $\omega(\mathsf{v}) \in \mathscr{F}(\mathcal{M})$ evaluated at p by setting

$$(\omega(\mathsf{v}))\,(p) := \langle\, \omega(p), \mathsf{v}(p)\, \rangle \qquad \forall\, p \in \mathcal{M}. \tag{1.3}$$

Formula (1.3) expresses the *evalutation* at p of a 1-form on a smooth vector field.

Example 1.6 (*The tangent space to implicit submanifolds of* $\mathbb{R}^n$)

Let

$$\mathcal{M} := \{x \in \mathbb{R}^n \,:\, F_i(x) = 0,\, i = 1, \ldots, k\}$$

be an implicit submanifold of $\mathbb{R}^n$. The tangent space to $\mathcal{M}$ at $p \in \mathcal{M}$ is

$$T_p\mathcal{M} := \left\{ y \in \mathbb{R}^n \,:\, \left\langle \left(\frac{\partial F_i}{\partial x_1}, \ldots, \frac{\partial F_i}{\partial x_n} \right)^{\top}, y \right\rangle = 0,\, i = 1, \ldots, k \right\}.$$

- Let $x := (x_1, x_2, x_3)^\top \in \mathbb{R}^3$. If $\mathcal{M} = \mathbb{S}^2$ and $F(x) := \langle x, x \rangle - 1$, we get

$$T_p\mathbb{S}^2 := \left\{ y \in \mathbb{R}^3 \,:\, \langle x, y \rangle = 0 \right\}, \qquad p \in \mathbb{S}^2.$$

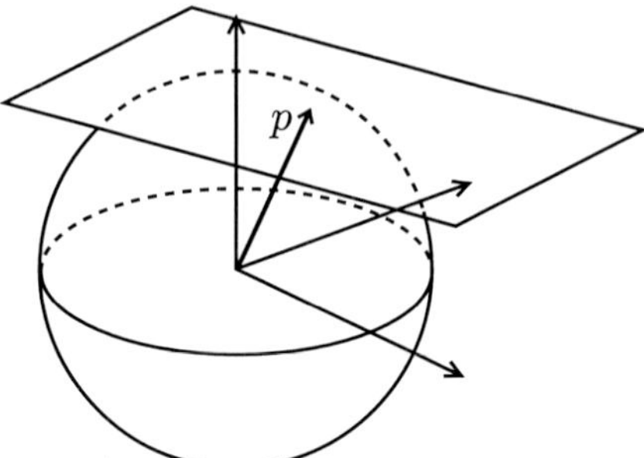

Fig. 4.3. A tangent space $T_p\mathbb{S}^2$.

- The tangent bundle $T\mathbb{S}^2$ is by definition

$$T\mathbb{S}^2 := \bigcup_{p \in \mathbb{S}^2} \{p\} \times T_p\mathbb{S}^2,$$

a four-dimensional smooth manifold that is not possible to visualize.
- A simpler example is the unit circle $\mathbb{S}^1$, whose tangent bundle is isomorphic to $\mathbb{S}^1 \times \mathbb{R}$. Geometrically, this is a cylinder of infinite height.

- With a function $F \in \mathscr{F}(\mathcal{M})$ we may associate its *differential* $\mathsf{d}F \in \Omega(\mathcal{M})$, which is a 1-form, that can be applied to vector fields on $\mathcal{M}$.

  (a) The notion of differential is used to associate to every vector field v on $\mathcal{M}$ a *derivation* on $\mathscr{F}(\mathcal{M})$. For $F \in \mathscr{F}(\mathcal{M})$ we define $\mathsf{v}[F] \in \mathscr{F}(\mathcal{M})$ by

$$\mathsf{v}[F] := \mathsf{d}F(\mathsf{v}), \tag{1.4}$$

  or, equivalently,

$$(\mathsf{v}[F])\,(p) := \langle\, \mathsf{d}F(p), \mathsf{v}(p)\, \rangle \qquad \forall\, p \in \mathcal{M}. \tag{1.5}$$

  Note that $(\mathsf{v}[F])\,(p)$ is the (directional) Lie derivative of F along the direction of v at p (to be formally defined later).

  (b) Saying that v is a *derivation* on $\mathscr{F}(\mathcal{M})$ means

$$\mathsf{v}[F\,G] = \mathsf{v}[F]\,G + F\,\mathsf{v}[G] \qquad \forall\, F, G \in \mathscr{F}(\mathcal{M}),$$

  which is a consequence of (1.5) and the *Leibniz rule* for differentials.

- It is a fundamental fact that every derivation on $\mathcal{M}$ corresponds to a unique vector field on $\mathcal{M}$ and that every derivation at p corresponds to a unique tangent vector at p. As a consequence, since the commutator of two derivations is a derivation, we may define the *Lie bracket* between two vector fields $\mathsf{v}_1, \mathsf{v}_2 \in \mathfrak{X}(\mathcal{M})$ as the vector field $[\mathsf{v}_1, \mathsf{v}_2]$ whose action on $F \in \mathscr{F}(\mathcal{M})$ is

$$[\mathsf{v}_1, \mathsf{v}_2][F] := \mathsf{v}_1[\mathsf{v}_2[F]] - \mathsf{v}_2[\mathsf{v}_1[F]]. \tag{1.6}$$

In this way $\mathfrak{X}(\mathcal{M})$ becomes the so called (infinite-dimensional) *Lie algebra of smooth vector fields*. In particular one has

$$[F\mathsf{v}_1, \mathsf{v}_2] = F[\mathsf{v}_1, \mathsf{v}_2] - \mathsf{v}_2[F]\,\mathsf{v}_1.$$

► We now provide a description in local coordinates. Let $\mathcal{U} \subset \mathcal{M}$ be a coordinate chart coordinatized by $x := (x_1, \ldots, x_n) : \mathcal{U} \to \mathbb{R}^n$ around a point $p \in \mathcal{M}$. Then a derivation on $\mathscr{F}(\mathcal{U})$ is completely determined once we know its effect on all coordinates x_i.

- The differential of $F \in \mathscr{F}(\mathcal{U})$ can be written as

  $$\mathrm{d}F = \sum_{i=1}^{n} \frac{\partial F}{\partial x_i} \mathrm{d}x_i,$$

  so that, in view of (1.5), we obtain

  $$\mathsf{v}[F(x)] = \sum_{i=1}^{n} \frac{\partial F}{\partial x_i} \mathsf{v}[x_i], \tag{1.7}$$

  which expresses the action of a smooth vector field on a smooth function.

- $T_p\mathcal{M}$ has basis $\{\partial/\partial x_1, \ldots, \partial/\partial x_n\}$, while $T_p^*\mathcal{M}$ has basis $\{\mathrm{d}x_1, \ldots, \mathrm{d}x_n\}$. In other words, a vector field on $\mathcal{M}$ has a local coordinate form

  $$\mathsf{v} = \sum_{i=1}^{n} f_i(x) \frac{\partial}{\partial x_i}, \qquad f_i \in \mathscr{F}(\mathcal{U}), \tag{1.8}$$

  while a 1-form on $\mathcal{M}$ can be represented as

  $$\omega = \sum_{i=1}^{n} g_i(x)\,\mathrm{d}x_i, \qquad g_i \in \mathscr{F}(\mathcal{U}).$$

  There holds the *duality pairing*

  $$\left\langle \mathrm{d}x_i, \frac{\partial}{\partial x_j} \right\rangle = \delta_{ij}, \qquad i, j = 1, \ldots, n.$$

- Remarkably, there exists a one-to-one correspondence between smooth vector fields on $\mathcal{U}$ and systems of ODEs on $\mathcal{U}$ of the form

$$\frac{\mathrm{d}x_i}{\mathrm{d}t} = f_i(x), \qquad i = 1, \dots, n, \tag{1.9}$$

where $f_i \in \mathscr{F}(\mathcal{U})$. Such result is sometimes called the "Fundamental Lie Theorem". Indeed, given a vector field v in the form (1.8) we have $f_i(x) := \mathsf{v}[x_i]$. Conversely, given the functions $f_i \in \mathscr{F}(\mathcal{U})$ we can define $\mathsf{v}[x_i] := f_i(x)$ and extend v to a derivation on $\mathscr{F}(\mathcal{U})$ by using (1.7). We list here some equivalent notations:

$$\mathsf{v}[x_i] \equiv \mathsf{v}_{x_i} \equiv \frac{\mathrm{d}x_i}{\mathrm{d}t} \equiv \dot{x}_i.$$

(a) Solutions to (1.9) define parametrized curves in $\mathcal{U}$, called *integral curves*, whose tangent vector at each point coincides with the value of v at that point.

(b) The local existence and uniqueness of integral curves of initial value problems of the form (1.9) is guaranteed by the "Picard-Lindelöf Theorem": given a smooth vector field v on $\mathcal{M}$ we can find for any $p \in \mathcal{M}$ a coordinate chart $\mathcal{U}$ of p, with coordinates $x := (x_1, \dots, x_n)$, an open subset $\mathcal{U}' \subseteq \mathcal{U}$ and $\varepsilon > 0$, such that the solution $(t, p) \mapsto x(t, p)$ of (1.9) is uniquely defined for $(t, p) \in \mathcal{I}_\varepsilon \times \mathcal{U}'$, where $\mathcal{I}_\varepsilon := \{t \in \mathbb{R} : |t| < \varepsilon\}$. The map

$$\Phi_t : \mathcal{I}_\varepsilon \times \mathcal{U}' \to \mathcal{U} \,:\, (t, p) \mapsto \Phi_t(p) := x(t, p)$$

is the *local flow* of v. For a fixed $t \in \mathcal{I}_\varepsilon$ the map Φ_t is a local diffeomorphism from $\mathcal{U}'$ to $\Phi_t(\mathcal{U}')$.

(c) Flows are *local one-parameter Lie groups of diffeomorphisms* for which there hold the properties

$$\Phi_{t+s}(p) = (\Phi_t \circ \Phi_s)(p) \qquad \forall\, t, s \in \mathcal{I}_\varepsilon,\ p \in \mathcal{U}', \tag{1.10}$$

$$\Phi_0(p) = p \qquad \forall\, p \in \mathcal{U}', \tag{1.11}$$

and

$$(\Phi_t \circ \Phi_{-t})(p) = p \qquad \forall\, t \in \mathcal{I}_\varepsilon,\ p \in \mathcal{U}'. \tag{1.12}$$

Let us recall that if $\mathcal{M}$ is compact, then all smooth vector fields on $\mathcal{M}$ are *complete*, so that Φ_t exists for all times t. We will often pretend that Φ_t is globally defined, i.e., the corresponding vector field is complete.

(d) Note that a solution $(t, p) \mapsto x(t, p)$ to (1.9) defines for each $p \in \mathcal{M}$ two different curves, the above mentioned *integral curve*,

$$\gamma_p := \left\{(t, x) \in \mathcal{I}_\varepsilon \times \mathcal{U}' \,:\, x(t, p) = \Phi_t(p),\ t \in \mathcal{I}_\varepsilon\right\},$$

and the *orbit* through p, which is the projection of γ_p onto $\mathcal{M}$:

$$\mathcal{O}_p := \{x \in \mathcal{U}' \, : \, x(t,p) = \Phi_t(p), \, t \in \mathcal{I}_\varepsilon\} .$$

Both curves γ_p and $\mathcal{O}_p$ are parametrized by the time t and oriented by the direction of time advance.

(e) One also says that the vector field v, given by (1.8) and expressed as a linear differential operator, is the *infinitesimal generator* of Φ_t. The component $f_i(x)$, given in (1.9), is the i-th component of

$$f(x) := \left.\frac{\mathrm{d}}{\mathrm{d}t}\right|_{t=0} \Phi_t(x).$$

One can prove that the following *Lie series* is valid:

$$(\Phi_t(x))_i = \exp\,(t\,\mathsf{v})\, x_i := \sum_{k=0}^{\infty} \frac{t^k}{k!} \mathsf{v}^k[x_i], \qquad i = 1, \ldots, n,$$

where $\mathsf{v}^k := \mathsf{v}\,\mathsf{v}^{k-1}$, v^0 being the identity.

(f) The "Picard-Lindelöf Theorem" implies another classical result, known as "Straightening Theorem": let v be a smooth vector field on $\mathcal{M}$ and suppose that $\mathsf{v}(p) \neq 0$, where $p \in \mathcal{M}$. Then there exist coordinates $(x_1, \ldots, x_n)$ on a neighborhood $\mathcal{U}$ of p such that the restriction of v to $\mathcal{U}$ is the first coordinate vector field, i.e., $\mathsf{v} = \partial/\partial x_1$.

Example 1.7 (*Rotations in* $\mathbb{R}^2$)

Consider the one-parameter Lie group of smooth diffeomorphisms on $\mathbb{R}^2$:

$$\begin{aligned} \Phi_t : \quad & [0, 2\,\pi) \times \mathbb{R}^2 \;\rightarrow\; \mathbb{R}^2, \\ & (t, (x_1, x_2)) \;\mapsto\; (\underline{x}_1, \underline{x}_2) := (x_1 \cos t + x_2 \sin t, -x_1 \sin t + x_2 \cos t). \end{aligned}$$

Note that Φ_t satisfies properties (1.10-1.12) and it defines planar rotations of angle t.

- The planar initial value problem that has Φ_t as flow is easily constructed. We have

$$\begin{cases} \dot{\underline{x}}_1 = f_1(\underline{x}_1, \underline{x}_2), \\ \dot{\underline{x}}_2 = f_2(\underline{x}_1, \underline{x}_2), \\ (\underline{x}_1(0, (x_1, x_2)), \underline{x}_2(0, (x_1, x_2)) = (x_1, x_2), \end{cases}$$

with

$$\begin{aligned} (f_1(\underline{x}_1, \underline{x}_2), f_2(\underline{x}_1, \underline{x}_2)) &:= \left.\frac{\mathrm{d}}{\mathrm{d}t}\right|_{t=0} (\underline{x}_1 \cos t + \underline{x}_2 \sin t, -\underline{x}_1 \sin t + \underline{x}_2 \cos t) \\ &= (\underline{x}_2, -\underline{x}_1). \end{aligned}$$

- The infinitesimal generator of the flow Φ_t is

$$\mathsf{v} := x_2 \frac{\partial}{\partial x_1} - x_1 \frac{\partial}{\partial x_2}.$$

- The Lie series of Φ_t reproduces the solution of the original initial value problem. One easily finds
$$\mathsf{v}^{2k+1}[x_1] = (-1)^k x_2, \qquad \mathsf{v}^{2k}[x_1] = (-1)^k x_1, \qquad k \in \mathbb{N}.$$
This leads to
$$\begin{aligned}(\Phi_t(x))_1 &= \exp(t\,\mathsf{v})\, x_1 = \sum_{k=0}^{\infty} \frac{t^k}{k!}\mathsf{v}^k\,[x_1] \\ &= x_1 \sum_{k=0}^{\infty} \frac{(-1)^k t^{2k}}{(2k)!} + x_2 \sum_{k=0}^{\infty} \frac{(-1)^k t^{2k+1}}{(2k+1)!} \\ &= x_1 \cos t + x_2 \sin t.\end{aligned}$$
Similarly for $(\Phi_t(x))_2 = -x_1 \sin t + x_2 \cos t$.

1.2.2 Maps between manifolds and submanifolds

► Let $\mathcal{M}, \mathcal{N}$ be two real smooth manifolds of dimension n and ℓ respectively.

- A continuous map $\varphi : \mathcal{M} \to \mathcal{N}$ is *differentiable* (resp. *smooth*) if for each point $p \in \mathcal{M}$, parametrized by a local coordinate function $\chi_\alpha : \mathcal{U}_\alpha \to \mathbb{R}^n$ there exists a coordinate chart $\mathcal{V}_\beta$ on $\mathcal{N}$ and a local coordinate function $\psi_\beta : \mathcal{V}_\beta \to \mathbb{R}^\ell$ with $\varphi(\chi_\alpha(\mathcal{U}_\alpha)) \subset \psi_\beta(\mathcal{V}_\beta)$ such that the composite function $\psi_\beta^{-1} \circ \varphi \circ \chi_\alpha : \mathcal{U}_\alpha \to \mathcal{V}_\beta$, called *local representation of* φ, is a differentiable (resp. *smooth*) function.

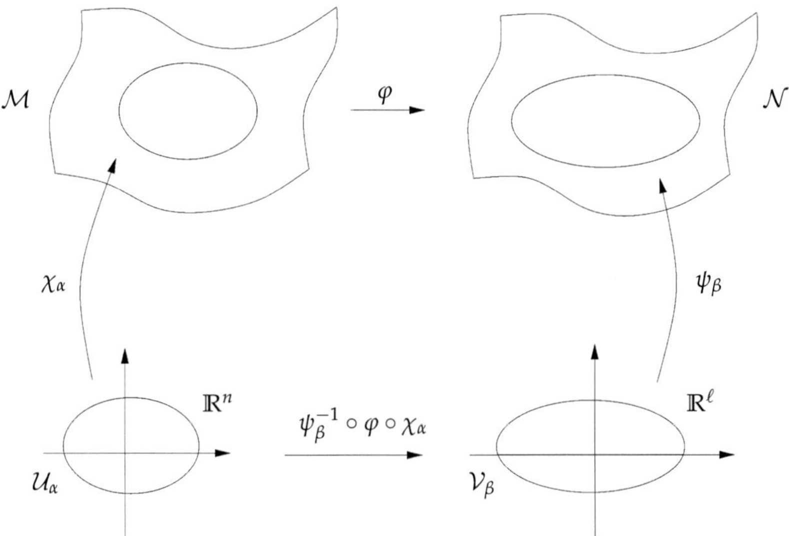

Fig. 4.4. A map $\varphi : \mathcal{M} \to \mathcal{N}$.

- The *differential* of a differentiable map $\varphi : \mathcal{M} \to \mathcal{N}$ at $p \in \mathcal{M}$ is the linear map
$$\mathrm{d}\varphi(p) : T_p\mathcal{M} \to T_{\varphi(p)}\mathcal{N}.$$
It is constructed as follows. Fix $p \in \mathcal{M}$ with coordinates $x := (x_1, \ldots, x_n)$. For $\mathsf{v} \in T_p\mathcal{M}$ choose a curve γ that maps an open interval $\mathcal{I}_\varepsilon := \{t \in \mathbb{R} : |t| < \varepsilon\}$ to $\mathcal{M}$ with $\gamma(0) = x$. The *velocity vector* is
$$\left.\frac{\mathrm{d}}{\mathrm{d}t}\right|_{t=0} \gamma(t) = \mathsf{v}(p).$$
Then $\mathrm{d}\varphi(p)(\mathsf{v})$ is the velocity vector at $t = 0$ of the curve $\varphi \circ \gamma : \mathcal{I}_\varepsilon \to \mathcal{N}$:
$$\mathrm{d}\varphi(p)(\mathsf{v}) := \left.\frac{\mathrm{d}}{\mathrm{d}t}\right|_{t=0} \varphi(\gamma(t)).$$

- The *rank* of a differentiable map $\varphi : \mathcal{M} \to \mathcal{N}$ at $p \in \mathcal{M}$ is the rank of $\mathrm{d}\varphi(p)$, i.e., the dimension of the image of $\mathrm{d}\varphi(p)$, that is at most $\min(n, \ell)$. One says that φ is of *maximal rank* on a subset $\mathcal{S} \subset \mathcal{M}$ if for each $p \in \mathcal{S}$ the rank of φ is $\min(n, \ell)$. When $n < \ell$, the best we can hope for is that $\mathrm{d}\varphi(p)$ is injective.

► Let us characterize some important smooth maps $\varphi : \mathcal{M} \to \mathcal{N}$.

- The map φ is a *diffeomorphism* if it is bijective and φ^{-1} is differentiable. In such a case the manifolds $\mathcal{M}$ and $\mathcal{N}$ are *diffeomorphic*.

- The map φ is an *immersion at a point* $p \in \mathcal{M}$ if $\mathrm{d}\varphi(p)$ is injective. It is an *immersion* if it is an immersion at all points $p \in \mathcal{M}$. Equivalently, φ is an immersion if
$$\mathrm{rank}(\mathrm{d}\varphi(p)) = n \qquad \forall\, p \in \mathcal{M}.$$

- The map φ is a *submersion at a point* $p \in \mathcal{M}$ if $\mathrm{d}\varphi(p)$ is surjective. Note that we necessarily have $n \geqslant \ell$. It is a *submersion* if it is a submersion at all points $p \in \mathcal{M}$. Equivalently, φ is a submersion if
$$\mathrm{rank}(\mathrm{d}\varphi(p)) = \ell \qquad \forall\, p \in \mathcal{M}.$$

- The map φ is an *embedding* if it is an (injective) immersion that is also a homeomorphism onto its image $\varphi(\mathcal{M})$.

► A point $q \in \mathcal{N}$ is called *regular value* of φ if $\mathrm{d}\varphi(p)$ is surjective for all $p \in \varphi^{-1}(q)$, i.e.,
$$\mathrm{rank}(\mathrm{d}\varphi(p)) = \ell \qquad \forall\, p \in \varphi^{-1}(q).$$
Otherwise it is called a *singular value*.

► Let $\varphi : \mathcal{M} \to \mathcal{N}$ be a smooth map.

- Assume that φ is an immersion. Then $\varphi(\mathcal{M})$ is an *immersed manifold* of $\mathcal{N}$. In general this image will not be a submanifold as a subset because an immersion is not necessarily injective. By requiring that the immersion map φ be injective one has an injective immersion and one can define an *immersed submanifold* of $\mathcal{N}$ as the image subset $\varphi(\mathcal{M})$ together with a topology and differential structure such that $\varphi(\mathcal{M})$ is a manifold and the injective immersion is a diffeomorphism.

- Assume that φ be an embedding. Then $\varphi(\mathcal{M})$ is an *embedded submanifold* (or *regular submanifold*) of $\mathcal{N}$. An embedded submanifold of $\mathcal{N}$ is such that the image $\varphi(\mathcal{M})$ with the subspace topology is homeomorphic to $\mathcal{M}$ under φ.

► The following classical results hold true:

- $\varphi : \mathcal{M} \to \mathcal{N}$ is a local diffeomorphism if and only if $\mathrm{d}\varphi(p)$ is an isomorphism for all $p \in \mathcal{M}$.

- "Maximal rank Theorem". Let $\varphi : \mathcal{M} \to \mathcal{N}$ be a smooth map with maximal rank at $p \in \mathcal{M}$, i.e., an immersion at p. Then there exist coordinates $(x_1, \dots, x_n)$ in a neighborhood of p and coordinates in a neighborhood of $\varphi(p)$ where φ has the following coordinate representation (*canonical immersion*):

$$\varphi(x_1, \dots, x_n) = \begin{cases} (x_1, \dots, x_n, 0, \dots, 0) & \text{if } \ell > n, \\ (x_1, \dots, x_\ell) & \text{if } \ell \leqslant n. \end{cases}$$

- "Regular value Theorem". Let $\varphi : \mathcal{M} \to \mathcal{N}$ be a smooth map with $\ell < n$. Let $q \in \mathcal{N}$ be a regular value of φ, with $\varphi^{-1}(q) \neq \emptyset$. Then:

  1. $\varphi^{-1}(q)$ is a smooth manifold of dimension $n - \ell$.
  2. The tangent space at $p \in \varphi^{-1}(q)$ is

$$T_p\, \varphi^{-1}(q) = \mathsf{Ker}(\mathrm{d}\varphi(p)).$$

- "Whitney Theorem". Any n-dimensional smooth manifold can be embedded in $\mathbb{R}^{2n}$. As a consequence any n-dimensional smooth manifold is diffeomorphic to a submanifold of $\mathbb{R}^{2n}$.

Example 1.8 (*The unit hypersphere* S^n)

The fact that the hypersphere S^n is a regular n-dimensional submanifold of $\mathbb{R}^{n+1}$ follows easily from the "Regular value Theorem".

- S^n is the regular level set $\varphi^{-1}(1)$ of the smooth function $\varphi : \mathbb{R}^{n+1} \to [0, \infty)$ defined by

$$\varphi(x) := \sum_{i=1}^{n+1} x_i^2.$$

Note that

$$\mathrm{d}\varphi = 2\sum_{i=1}^{n+1} x_i \, \mathrm{d}x_i$$

vanishes only at $x = 0$.

- Moreover, given $p \in \mathbb{S}^n$, with coordinates $(x_1, \ldots, x_{n+1})$, we have

$$T_p\varphi^{-1}(1) = T_p\mathbb{S}^n = \mathsf{Ker}(\mathrm{d}\varphi(p)) = \left\{ y \in \mathbb{R}^{n+1} \, : \, \sum_{i=1}^{n+1} x_i \, y_i = 0 \right\},$$

which is a hyperplane tangent to $\mathbb{S}^n$ at p.

► We finally recall the notion of connectedness.

- In general, a topological space is *connected* if it cannot be decomposed into the disjoint union of two or more non-empty open subsets.
- The maximal connected subsets (ordered by inclusion) of a non-empty topological space are called the *connected components* of the space. The components of any topological space form a partition: they are disjoint, non-empty, and their union is the whole space. Every component is a closed subset of the original space. It follows that if their number is finite then each component is also an open subset.
- Since any smooth real n-dimensional manifold locally looks like $\mathbb{R}^n$, it is not difficult to prove that any connected smooth manifold is *pathwise connected*, meaning that there exists a smooth curve joining any pair of points.

1.2.3 Distributions

► Instead of having a vector at every point of a manifold $\mathcal{M}$, as in the case of a vector field on $\mathcal{M}$, one may have a one-dimensional subspace of the tangent space to $\mathcal{M}$, at every point of $\mathcal{M}$. This is called a *1-dimensional smooth distribution* on $\mathcal{M}$.

► An *r-dimensional smooth distribution* $\mathcal{D}$ on $\mathcal{M}$ $(r < n)$ is a choice of an r-dimensional linear subspace $\mathcal{D}(p)$ of $T_p\mathcal{M}$ for every $p \in \mathcal{M}$, so that there exist smooth vector fields $\mathsf{v}_1, \ldots, \mathsf{v}_r$ on a open neighborhood $\mathcal{U}$ of p, such that

$$\mathcal{D}(p) = \mathsf{span}\{\mathsf{v}_1(p), \ldots, \mathsf{v}_r(p)\} \qquad \forall \, p \in \mathcal{U}.$$

If the dimension of $\mathcal{D}$ varies with p then the distribution is called *singular*.

- The notion of integral curves of vector fields can be generalized to the case of a k-dimensional distribution: an r-dimensional (connected) immersed submanifold $\mathcal{N}$ of $\mathcal{M}$ is called an *integral manifold* of $\mathcal{D}$ if $T_p\mathcal{N} = \mathcal{D}(p)$ for any $p \in \mathcal{N}$.

- In contrast to the case of integral curves, integral manifolds need not to exist in general, even locally.
  (a) We say that a vector field v on $\mathcal{U} \subset \mathcal{M}$ is *adapted* to $\mathcal{D}$ on $\mathcal{U}$ if $\mathsf{v}(p) \in \mathcal{D}(p)$ for every $p \in \mathcal{U}$.
  (b) The obstruction to existence of integral manifolds reads: if v_1 and v_2 are adapted to $\mathcal{D}$ on $\mathcal{U}$ then, in general, it is not true that $[\mathsf{v}_1, \mathsf{v}_2]$ is also adapted to $\mathcal{D}$ on $\mathcal{U}$. If this happens for any v_1 and v_2 on $\mathcal{U}$ one says that $\mathcal{D}$ is *integrable (in the sense of Frobenius) on* $\mathcal{U}$.

► The "Frobenius Theorem" says that the above obstruction to the existence of integral manifolds is the only one.

- "Frobenius Theorem" (first formulation). Let $\mathcal{D}$ be a smooth r-dimensional distribution on $\mathcal{M}$. If $\mathcal{D}$ is integrable (in the sense of Frobenius) then there exists through any point $p \in \mathcal{M}$ a unique maximal integral manifold for $\mathcal{D}$.
- "Frobenius Theorem" (second formulation). Let $\mathcal{D}$ be an integrable (in the sense of Frobenius) smooth r-dimensional distribution on $\mathcal{M}$. Then we can choose coordinates $(x_1, \ldots, x_n)$ on a neighborhood $\mathcal{U}$ of any $p \in \mathcal{M}$ such that

$$\mathcal{D}(q) = \mathsf{span}\left\{ \left(\frac{\partial}{\partial x_1}\right)(q), \ldots, \left(\frac{\partial}{\partial x_r}\right)(q) \right\}$$

  for any $q \in \mathcal{U}$. In terms of these coordinates the integral manifold of $\mathcal{D}$ restricted to $\mathcal{U}$ and through p is given by the connected component of the set $\{q \in \mathcal{U} \,:\, x_i(q) = x_i(p),\, i = r+1, \ldots, n\}$ that contains p.

Note that the first formulation is the generalization of "Picard-Lindelöf Theorem" while the second formulation is a generalization of the "Straightening Theorem".

► Remarks:

- A collection of smooth vector fields $\mathsf{v}_1, \ldots, \mathsf{v}_r$ defined in a chart $\mathcal{U}$ of $p \in \mathcal{M}$ is *involutive* if there exist functions $c_{ij}^k \in \mathscr{F}(\mathcal{U})$, $i, j, k = 1, \ldots, r$, such that

$$[\mathsf{v}_i, \mathsf{v}_j] = \sum_{k=1}^{r} c_{ij}^k(x)\, \mathsf{v}_k,$$

  for all $i, j = 1, \ldots, r$. In terms of this definition one can reformulate "Frobenius Theorem" by claiming that a collection of smooth vector fields $\mathsf{v}_1, \ldots, \mathsf{v}_r$ is *integrable* (in the sense of Frobenius) (i.e., through every point $p \in \mathcal{M}$ there passes an integral submanifold) if and only if it is involutive.

- "Frobenius Theorem" implies that the maximal integral manifolds of an integrable (in the sense of Frobenius) distribution on $\mathcal{M}$ form the leaves of a *foliation* on $\mathcal{M}$.

  (a) Roughly speaking, a foliation of a smooth manifold $\mathcal{M}$ looks locally like a decomposition of $\mathcal{M}$ into the union of "parallel" submanifolds of smaller dimension. More formally, an r-dimensional *foliation* (or *stratification*) of an n-dimensional smooth manifold $\mathcal{M}$ is a countable collection of subsets $\mathcal{M}_\alpha$ of $\mathcal{M}$ (called *leaves* or *strata*) such that:

     1. There holds
     $$\bigcup_\alpha \mathcal{M}_\alpha = \mathcal{M}, \qquad \mathcal{M}_\alpha \cap \mathcal{M}_\beta = \emptyset, \quad \alpha \neq \beta.$$
     2. Each leaf $\mathcal{M}_\alpha$ is pathwise connected (i.e., if $p, q \in \mathcal{M}_\alpha$ then there exists a continuous curve $\gamma : [0,1] \to \mathcal{M}_\alpha$ such that $\gamma(0) = p, \gamma(1) = q$).
     3. For each point $p \in \mathcal{M}$ there exists a local chart $\mathcal{U}$ containing p with coordinates $(x_1, \ldots, x_n) : \mathcal{U} \to \mathbb{R}^n$ such that the connected components of the intersections of the leaves with $\mathcal{U}$ are the level sets of $(x_{r+1}, \ldots, x_n) : \mathcal{U} \to \mathbb{R}^{n-r}$.

     The coordinates $(x_1, \ldots, x_r)$ provide local coordinates on $\mathcal{M}_\alpha$, which are therefore images of injective immersions. In particular, the leaves have well defined r-dimensional tangent spaces at each point. Therefore any foliation of dimension r defines an r-dimensional distribution.

  (b) There is a close relationship between foliations and integrability (in the sense of Frobenius) of a distribution. For instance, given a vector field on $\mathcal{M}$ that is never zero, its integral curves will give a one-dimensional foliation. This observation is generalized by "Frobenius Theorem" saying that the necessary and sufficient conditions for a distribution to be tangent to the leaves of a foliation, is that the set of vector fields tangent to the distribution are closed under the Lie bracket.

1.2.4 k-forms

► In the formulation of Hamiltonian mechanics on smooth manifolds we will be interested in differential 2-forms and their dual version, namely 2-vector fields (also called *bivector fields*). In order to understand the main features of these objects in a general geometric framework we give a review of k-forms and k-vector fields.

► The natural generalization of a 1-form on $\mathcal{M}$ is given by a differential k-form on $\mathcal{M}$, whose local definition is nothing but the definition of a differential k-form on a subset $\mathcal{U}$ of $\mathbb{R}^n$. To fix some notation we start with a local description. Let $\mathcal{U} \subset \mathcal{M}$ be a coordinate chart coordinatized by $x := (x_1, \ldots, x_n)$ around a point $p \in \mathcal{M}$.

- The set of all differential k-forms on $\mathcal{U}$ is a vector space denoted by $\Omega^k(\mathcal{U})$, with

$$\operatorname{dim}\left(\Omega^k(\mathcal{U})\right) = \frac{n!}{k!(n-k)!}.$$

- A *differential k-form* on $\mathcal{U}$ is specified by $n!/k!/(n-k)!$ functions $f_{\ell_1\cdots\ell_k} \in \mathscr{F}(\mathcal{U})$:

$$\omega = \sum_{1\leqslant\ell_1<\cdots<\ell_k\leqslant n} f_{\ell_1\cdots\ell_k}(x)\, \mathrm{d}x_{\ell_1} \wedge \cdots \wedge \mathrm{d}x_{\ell_k}, \tag{1.13}$$

where $\wedge$ denotes the exterior product between differentials. Note that ω can be evaluated on k vector fields to give a smooth function.

- More generally, the *exterior product* of two differential forms $\omega \in \Omega^k(\mathcal{U})$, $\omega' \in \Omega^\ell(\mathcal{U})$ is the differential $(k+\ell)$-form defined by

$$(\omega \wedge \omega')(\mathsf{v}_1, \dots, \mathsf{v}_{k+\ell}) := \sum_{\sigma\in\pi} \nu(\sigma)\, \omega(\mathsf{v}_{\sigma_1}, \dots, \mathsf{v}_{\sigma_k})\, \omega'(\mathsf{v}_{\sigma_{k+1}}, \dots, \mathsf{v}_{\sigma_{k+\ell}}),$$

where $\mathsf{v}_1, \dots, \mathsf{v}_{k+\ell}$ are $k+\ell$ vector fields, $\sigma := (\sigma_1, \dots, \sigma_{k+\ell})$, π is the set of all permutations of $(1, \dots, k+\ell)$ and $\nu = \pm 1$ according to the parity of the permutation. This operation is associative, distributive and graded commutative, which means that

$$\omega \wedge \omega' = (-1)^{k\ell} \omega' \wedge \omega. \tag{1.14}$$

- The *differential* (or *exterior derivative*) of the differential k-form (1.13) is the differential $(k+1)$-form

$$\mathrm{d}\omega := \sum_{i=1}^{n} \sum_{1\leqslant\ell_1<\cdots<\ell_k\leqslant n} \frac{\partial f_{\ell_1\cdots\ell_k}}{\partial x_i}\, \mathrm{d}x_i \wedge \mathrm{d}x_{\ell_1} \wedge \cdots \wedge \mathrm{d}x_{\ell_k}. \tag{1.15}$$

It has the following properties:

1. *Linearity*. There holds:

$$\mathrm{d}(\lambda_1 \omega + \lambda_2 \omega') = \lambda_1\, \mathrm{d}\omega + \lambda_2\, \mathrm{d}\omega', \qquad \omega, \omega' \in \Omega^k(\mathcal{U}),\ \lambda_1, \lambda_2 \in \mathbb{R}.$$

2. *Anti-derivation* (graded Leibniz rule). There holds:

$$\mathrm{d}(\omega \wedge \omega') = \mathrm{d}\omega \wedge \omega' + (-1)^k \omega \wedge \mathrm{d}\omega', \qquad \omega \in \Omega^k(\mathcal{U}),\ \omega' \in \Omega^\ell(\mathcal{U}).$$

3. *Closure*. There holds:

$$\mathrm{d}(\mathrm{d}\omega) = 0, \qquad \omega \in \Omega^k(\mathcal{U}). \tag{1.16}$$

► We now provide a coordinate-free presentation and we take into account that every coordinate-dependent representation in a neighborhood $\mathcal{U}$ is given by the previous formulas, e.g., (1.13) and (1.15).

- For $k \in \mathbb{N}$ we denote the $\mathscr{F}(\mathcal{M})$-module of k-forms on $\mathcal{M}$ by $\Omega^k(\mathcal{M})$. In particular, $\Omega^0(\mathcal{M}) = \mathscr{F}(\mathcal{M})$ and $\Omega^1(\mathcal{M}) = \Omega(\mathcal{M})$.

- We let
$$\Omega^*(\mathcal{M}) := \bigoplus_{k=0}^{n} \Omega^k(\mathcal{M}), \qquad \mathsf{dim}\left(\Omega^k(\mathcal{M})\right) = \frac{n!}{k!(n-k)!}.$$
Any element of $\Omega^*(\mathcal{M})$ is a *differential form.*

- There is a $\mathscr{F}(\mathcal{M})$-bilinear map
$$\wedge : \Omega^*(\mathcal{M}) \times \Omega^*(\mathcal{M}) \to \Omega^*(\mathcal{M}),$$
which associates to two differential forms ω and ω' their *exterior product* $\omega \wedge \omega'$. This operation makes $\Omega^*(\mathcal{M})$ into a graded associative algebra over $\mathscr{F}(\mathcal{M})$ called *Grassmann algebra* (or *exterior algebra*) of $\mathcal{M}$. In particular formula (1.14) holds true.

- A *differential k-form* is an element of $\Omega^k(\mathcal{M})$, which is the set of all k-linear alternating maps
$$\omega : \underbrace{T_p\mathcal{M} \times \cdots \times T_p\mathcal{M}}_{k \text{ times}} \to \mathscr{F}(\mathcal{M}).$$
In other words, a differential k-form is a smooth section of the k-th exterior power of the cotangent bundle of $\mathcal{M}$. Then the basic requirements are:

  1. *Multilinearity*. For any k smooth vectors fields $(\mathsf{v}_1, \ldots, \mathsf{v}_k)$, a smooth vector field w and two scalars $\lambda_1, \lambda_2 \in \mathbb{R}$, there holds:
  $$\omega(\lambda_1 \mathsf{v}_1 + \lambda_2 \mathsf{w}, \mathsf{v}_2, \ldots, \mathsf{v}_k) = \lambda_1 \omega(\mathsf{v}_1, \mathsf{v}_2, \ldots, \mathsf{v}_k) + \lambda_2 \omega(\mathsf{w}, \mathsf{v}_2, \ldots, \mathsf{v}_k).$$
  2. *Skew-symmetry*. For any k smooth vector fields $(\mathsf{v}_1, \ldots, \mathsf{v}_k)$ there holds:
  $$\omega(\mathsf{v}_1, \ldots, \mathsf{v}_k) = (-1)^{\nu} \omega(\mathsf{v}_{i_1}, \ldots, \mathsf{v}_{i_k}),$$
  where $\nu = \pm 1$ according to the parity of the permutation $(i_1, \ldots, i_k)$ of $(1, \ldots, k)$.

- The *differential* (or *exterior derivative*) is a linear map $\mathsf{d} : \Omega^*(\mathcal{M}) \to \Omega^*(\mathcal{M})$ which maps k-forms to $(k+1)$-forms according to the following formula:
$$\begin{aligned} \mathsf{d}\omega(\mathsf{v}_0, \ldots, \mathsf{v}_k) \quad := \quad & \sum_{i=0}^{k} (-1)^i \mathsf{v}_i \left[\omega\left(\mathsf{v}_0, \ldots, \not{\mathsf{v}_i}, \ldots, \mathsf{v}_k\right)\right] \\ & + \sum_{i<j} (-1)^{i+j} \omega\left(\left[\mathsf{v}_i, \mathsf{v}_j\right], \mathsf{v}_0, \ldots, \not{\mathsf{v}_i}, \ldots, \not{\mathsf{v}_j}, \ldots, \mathsf{v}_k\right). \end{aligned} \tag{1.17}$$
A coordinate dependent local definition is (1.15).

- In general, a *complex* is defined as a sequence of vector spaces and linear maps between successive spaces with the following two basic properties:

  1. The composition of any pair of successive maps is identically zero (*closure property*).
  2. The kernel of one of the linear maps contains the image of the preceding map. If this containment is equality then the complex satisfies the *exactness property*.

- The differential d is used to define the *de Rham complex*:

$$\mathbb{R} \longrightarrow \Omega^0(\mathcal{M}) \xrightarrow{\mathsf{d}} \Omega^1(\mathcal{M}) \xrightarrow{\mathsf{d}} \cdots \xrightarrow{\mathsf{d}} \Omega^{n-1}(\mathcal{M}) \xrightarrow{\mathsf{d}} \Omega^n(\mathcal{M}) \longrightarrow 0.$$

  Note that the initial map $\mathbb{R} \longrightarrow \Omega^0(\mathcal{M})$ maps a constant $c \in \mathbb{R}$ to a constant function (0-form). The last transition $\Omega^n(\mathcal{M}) \longrightarrow 0$ simply means that there are no nonzero $(n+1)$-forms since $\mathcal{M}$ is n-dimensional.

  (a) The closure property of the complex is equivalent to the closure property of the differential:
  $$\mathsf{d} \circ \mathsf{d} = \mathsf{d}^2 = 0.$$

  (b) The exactness property means that a *closed differential form* $\omega \in \Omega^k(\mathcal{M})$, i.e., $\mathsf{d}\omega = 0$, is necessarily an *exact differential form*, meaning that there exists a $(k-1)$-form $\omega' \in \Omega^{k-1}(\mathcal{M})$ such that
  $$\omega = \mathsf{d}\omega'.$$

  (c) Clearly, any exact form is closed, but the converse is in general not true. Therefore the de Rham complex is not in general exact. However, on the local side and for special types of domains $\mathcal{U} \subset \mathbb{R}^n$ we do have exactness of the de Rham complex. This famous result, known as "Poincaré Lemma", holds for star-shaped domains $\mathcal{U}$ (or simply connected domains), where "star-shaped" means that whenever $x \in \mathcal{U}$, so is the entire line segment joining x to the origin: $\{\lambda x : \lambda \in [0,1]\} \subset \mathcal{U}$.

Example 1.9 (*A differential 2-form on* $\mathcal{M}$)

A differential 2-form on $\mathcal{M}$ admits the following local expression in a chart $\mathcal{U}$ (cf. formula (1.13)):

$$\omega = \sum_{1 \leqslant i < j \leqslant n} f_{ij}(x)\, \mathsf{d}x_i \wedge \mathsf{d}x_j,$$

where $f_{ij} \in \mathscr{F}(\mathcal{U})$. Its exterior derivative gives a 3-form on $\mathcal{M}$ whose local expression can be computed by using (1.15). A more intrinsic definition is (see (1.17)):

$$\mathsf{d}\omega(\mathsf{v}_1, \mathsf{v}_2, \mathsf{v}_3) := \mathsf{v}_1[\omega(\mathsf{v}_2, \mathsf{v}_3)] + \omega(\mathsf{v}_1, [\mathsf{v}_2, \mathsf{v}_3]) + \circlearrowleft (\mathsf{v}_1, \mathsf{v}_2, \mathsf{v}_3). \tag{1.18}$$

Hereafter we introduce the symbol $\circlearrowleft (\mathsf{v}_1, \mathsf{v}_2, \mathsf{v}_3)$ to denote any cyclic permutation of $(\mathsf{v}_1, \mathsf{v}_2, \mathsf{v}_3)$.

► The fact that a 1-form can be evaluated on a vector field to produce an element of $\mathscr{F}(\mathcal{M})$ (see (1.3)) generalizes in two ways to a k-form ω, where $k > 1$.

1. We can evaluate $\omega \in \Omega^k(\mathcal{M})$ on k vector fields $\mathsf{v}_1, \ldots, \mathsf{v}_k$ obtaining
$$\omega(\mathsf{v}_1, \ldots, \mathsf{v}_k) \in \mathscr{F}(\mathcal{M}).$$
From this point of view a k-form is an $\mathscr{F}(\mathcal{M})$-k-linear map on $\mathfrak{X}(\mathcal{M})$ with values in $\mathscr{F}(\mathcal{M})$.

2. We can insert one vector field v as the first argument to $\omega \in \Omega^k(\mathcal{M})$ yielding a $(k-1)$-form that is denoted by $\mathsf{v} \lrcorner \omega$ and called *interior product* (or *contraction*):
$$\mathsf{v} \lrcorner \omega \in \Omega^{k-1}(\mathcal{M}).$$
From this point of view a k-form is an $\mathscr{F}(\mathcal{M})$-linear map $\mathfrak{X}(\mathcal{M}) \to \Omega^{k-1}(\mathcal{M})$. The definition of the map $\mathsf{v} \lrcorner$ can be extended to all of $\Omega^*(\mathcal{M})$ by defining $\mathsf{v} \lrcorner \omega = 0$ for all 0-forms on $\mathcal{M}$.

Example 1.10 (*A contraction of a 3-form with a vector field*)

On $\mathbb{R}^3$ define a 3-form
$$\omega := \mathrm{d}x_1 \wedge \mathrm{d}x_2 \wedge \mathrm{d}x_3,$$
and a vector field
$$\mathsf{v} := x_1 \frac{\partial}{\partial x_1} + x_2 \frac{\partial}{\partial x_2}.$$
Then
$$\mathsf{v} \lrcorner \omega = x_1 \, \mathrm{d}x_2 \wedge \mathrm{d}x_3 - x_2 \, \mathrm{d}x_1 \wedge \mathrm{d}x_3.$$

► Let $\varphi : \mathcal{M} \to \mathcal{N}$ be a smooth map between two smooth manifolds. Since $\mathrm{d}\varphi(p)$, $p \in \mathcal{M}$, is a linear isomorphism between $T_p\mathcal{M}$ and $T_{\varphi(p)}\mathcal{N}$ we can dualize this map, thus getting a linear map on cotangent spaces going in the opposite direction. More precisely, we can associate to $\omega \in \Omega^k(\mathcal{N})$ a differential k-form on $\Omega^k(\mathcal{M})$, denoted by $\varphi^\star(\omega)$ and called *pull-back* of ω, by setting
$$\big((\varphi^\star\omega)(\mathsf{v}_1, \ldots, \mathsf{v}_k)\big)(p) := \big(\omega(\mathrm{d}\varphi(p)(\mathsf{v}_1), \ldots, \mathrm{d}\varphi(p)(\mathsf{v}_k))\big)(\varphi(p)),$$
where $p \in \mathcal{M}$ and $\mathsf{v}_1, \ldots, \mathsf{v}_k \in T_p\mathcal{M}$.

- In the particular case $k = 0$, i.e., ω is a smooth function, say $F \in \mathscr{F}(\mathcal{M})$, we simply recover the pull-back of a function, i.e., $\varphi^\star F := F \circ \varphi$.
- Note that the pull-back of ω under φ is a linear operator from $\Omega^k(\mathcal{N})$ to $\Omega^k(\mathcal{M})$, and the word "pull-back" reminds one of the fact that the direction of the arrow is reversed compared to the arrow of the map $\varphi : \mathcal{M} \to \mathcal{N}$, i.e., $\varphi^\star\omega : \Omega^k(\mathcal{N}) \to \Omega^k(\mathcal{M})$.
- It can be proved that the pull-back commutes with the differential, i.e.,
$$\varphi^\star(\mathrm{d}\omega) = \mathrm{d}(\varphi^\star\omega). \tag{1.19}$$

Example 1.11 (*Pull-back of a 2-form on* $\mathbb{R}^2$)

Consider the map

$$\varphi : \mathbb{R}^2 \to \mathbb{R}^2 : (x_1, x_2) \mapsto (\widetilde{x}_1, \widetilde{x}_2) := \left(\frac{1}{2}\left(x_1^2 - x_2^2\right), x_1\, x_2\right).$$

Then

$$\begin{aligned} \mathrm{d}\widetilde{x}_1 \wedge \mathrm{d}\widetilde{x}_2 &= (x_1\,\mathrm{d}x_1 - x_2\,\mathrm{d}x_2) \wedge (x_2\,\mathrm{d}x_1 + x_1\,\mathrm{d}x_2) \\ &= \left(x_1^2 + x_2^2\right) \mathrm{d}x_1 \wedge \mathrm{d}x_2. \end{aligned}$$

Therefore $\left(x_1^2 + x_2^2\right) \mathrm{d}x_1 \wedge \mathrm{d}x_2$ is the pull-back of $\mathrm{d}\widetilde{x}_1 \wedge \mathrm{d}\widetilde{x}_2$ via φ:

$$\varphi^\star(\mathrm{d}\widetilde{x}_1 \wedge \mathrm{d}\widetilde{x}_2) = \left(x_1^2 + x_2^2\right) \mathrm{d}x_1 \wedge \mathrm{d}x_2.$$

► Remarks:

- The space $\Omega^n(\mathcal{M})$ of n-forms on $\mathcal{M}$ (that is n-dimensional) is one-dimensional. Thus all n-forms on $\mathcal{M}$ can be locally written in a coordinate chart $\mathcal{U}$ as

$$\omega = f(x)\,\mathrm{d}x_1 \wedge \cdots \wedge \mathrm{d}x_n, \tag{1.20}$$

  for some $f \in \mathscr{F}(\mathcal{U})$. If f is thought of as a measure on $\mathcal{U}$, then ω can be used to define the measure of a compact subset $\mathcal{U}' \subset \mathcal{U}$,

$$\mathsf{Vol}\,(\mathcal{U}') := \int_{\mathcal{U}'} \omega.$$

  Indeed, ω is also called *measure form* (*volume form* if $f \equiv 1$).

- Given a n-form ω as in (1.20) we say that a diffeomorphism $\varphi : \mathcal{M} \to \mathcal{M}$ is *volume preserving* if $\varphi^\star\omega = \omega$ when $f \equiv 1$ and *measure preserving* if $\varphi^\star\omega = \omega$ with $f \neq 1$. The distinction is slight since φ depends on the Jacobian of any coordinate transformation.

1.2.5 *k-vector fields*

► We now introduce k-vector fields on $\mathcal{M}$ as those objects that are dual to k-forms.

- For $k \in \mathbb{N}$ we denote the $\mathscr{F}(\mathcal{M})$-module of k-vector fields on $\mathcal{M}$ by $\mathfrak{X}^k(\mathcal{M})$. In particular, $\mathfrak{X}^0(\mathcal{M}) = \mathscr{F}(\mathcal{M})$ and $\mathfrak{X}^1(\mathcal{M}) = \mathfrak{X}(\mathcal{M})$.

- We let

$$\mathfrak{X}^*(\mathcal{M}) := \bigoplus_{k=0}^{n} \mathfrak{X}^k(\mathcal{M}), \qquad \mathsf{dim}\left(\mathfrak{X}^k(\mathcal{M})\right) = \frac{n!}{k!(n-k)!}.$$

  An element of $\mathfrak{X}^*(\mathcal{M})$ is a *multi-vector field*.

- A *k-vector field* is an element of $\mathfrak{X}^k(\mathcal{M})$, which is the set of all k-linear alternating maps
$$\mathsf{V} : \underbrace{T_p^*\mathcal{M} \times \cdots \times T_p^*\mathcal{M}}_{k \text{ times}} \to \mathscr{F}(\mathcal{M}).$$
In other words, a k-vector field is a smooth section of the k-th exterior power of the tangent bundle of $\mathcal{M}$. Therefore V defines a skew-symmetric k-derivation.

- As in the case of differential forms we can define an *exterior product*:
$$\wedge : \mathfrak{X}^*(\mathcal{M}) \times \mathfrak{X}^*(\mathcal{M}) \to \mathfrak{X}^*(\mathcal{M}),$$
which makes $\mathfrak{X}^*(\mathcal{M})$ into a graded associative algebra.

- We can evaluate any k-form ω on any k-vector field $\mathsf{V} := \mathsf{v}_1 \wedge \cdots \wedge \mathsf{v}_k$ by letting $\omega(\mathsf{V}) := \omega(\mathsf{v}_1, \ldots, \mathsf{v}_k)$.

- Let V be a k-vector field on $\mathcal{M}$. Then the value of the corresponding skew-symmetric k-derivation on k functions $F_1, \ldots, F_k \in \mathscr{F}(\mathcal{M})$ is denoted by
$$\mathsf{V}\left[F_1, \ldots, F_k\right] := \left(\mathrm{d}F_1 \wedge \cdots \wedge \mathrm{d}F_k\right)(\mathsf{V}). \tag{1.21}$$
Note that (1.21) is the generalization of (1.4) to k-vector fields.

- As in the case of smooth vector fields (cf. formula (1.7)) we have that a k-vector field V is completely specified on a coordinate neighborhood $\mathcal{U}$ once it is known on all k-tuples $(x_{i_1}, \ldots, x_{i_k})$, with $1 \leqslant i_1 < i_2 < \cdots < i_k \leqslant n$, where the k-tuples are taken from any chosen system of coordinates $(x_1, \ldots, x_n)$ on $\mathcal{U}$. Explicitly (1.7) is generalized to
$$\mathsf{V}\left[F_1, \ldots, F_k\right] = \sum_{i_1, \ldots, i_k = 1}^{n} \frac{\partial F_1}{\partial x_{i_1}} \cdots \frac{\partial F_k}{\partial x_{i_k}} \mathsf{V}[x_{i_1}, \ldots, x_{i_k}].$$

Example 1.12 (*A 2-vector field on* $\mathcal{M}$)

A 2-vector field on $\mathcal{M}$ admits the following local expression in a chart $\mathcal{U}$:
$$\mathsf{V} = \sum_{1 \leqslant i < j \leqslant n} f_{ij}(x) \frac{\partial}{\partial x_i} \wedge \frac{\partial}{\partial x_j},$$
where $f_{ij} \in \mathscr{F}(\mathcal{U})$.

Example 1.13 (*A 2-vector field on* $\mathbb{R}^3$)

Let $\mathcal{U}$ be a local chart on a three-dimensional manifold $\mathcal{M}$.

- A basis of $\Omega^2(\mathcal{U})$ is
$$\{\mathrm{d}x_2 \wedge \mathrm{d}x_3, \mathrm{d}x_3 \wedge \mathrm{d}x_1, \mathrm{d}x_1 \wedge \mathrm{d}x_2\}.$$

Indeed, a 2-form on $\mathcal{M}$ is expressed as

$$\omega = f_1(x)\,\mathrm{d}x_2 \wedge \mathrm{d}x_3 + f_2(x)\,\mathrm{d}x_3 \wedge \mathrm{d}x_1 + f_3(x)\,\mathrm{d}x_1 \wedge \mathrm{d}x_2,$$

for some functions $f_i \in \mathscr{F}(\mathcal{U})$, $i = 1,2,3$.

- A basis of $\mathfrak{X}^2(\mathcal{U})$ is

$$\left\{ \frac{\partial}{\partial x_2} \wedge \frac{\partial}{\partial x_3}, \frac{\partial}{\partial x_3} \wedge \frac{\partial}{\partial x_1}, \frac{\partial}{\partial x_1} \wedge \frac{\partial}{\partial x_2} \right\}.$$

A 2-vector field on $\mathcal{M}$ is expressed as

$$\mathsf{V} = g_1(x)\,\frac{\partial}{\partial x_2} \wedge \frac{\partial}{\partial x_3} + g_2(x)\,\frac{\partial}{\partial x_3} \wedge \frac{\partial}{\partial x_1} + g_3(x)\,\frac{\partial}{\partial x_1} \wedge \frac{\partial}{\partial x_2},$$

for some functions $g_i \in \mathscr{F}(\mathcal{U})$, $i = 1,2,3$.

1.2.6 Lie derivatives

► In general, Lie derivatives are important derivation operations which measure how a given object, as a k-form or a k-vector field, changes in the direction of a given vector field v. Here are the main definitions (in the case of k-vector fields we consider only $k = 1,2$).

- The *Lie derivative of a k-form* $\omega \in \Omega^k(\mathcal{M})$ along a smooth vector field $\mathsf{v} \in \mathfrak{X}(\mathcal{M})$ is the k-form whose value on $\mathsf{v}_1, \ldots, \mathsf{v}_k \in \mathfrak{X}(\mathcal{M})$ is given by

$$\mathfrak{L}_{\mathsf{v}}\omega(\mathsf{v}_1,\ldots,\mathsf{v}_k) := \mathsf{v}[\omega(\mathsf{v}_1,\ldots,\mathsf{v}_k)] - \sum_{i=1}^{k} \omega\,(\mathsf{v}_1,\ldots,[\mathsf{v},\mathsf{v}_i],\ldots,\mathsf{v}_k)\,. \tag{1.22}$$

A more geometric definition is the following. The *Lie derivative of a k-form* $\omega \in \Omega^k(\mathcal{M})$ along a smooth vector field $\mathsf{v} \in \mathfrak{X}(\mathcal{M})$ is

$$\mathfrak{L}_{\mathsf{v}}\omega := \left.\frac{\mathsf{d}}{\mathsf{d}t}\right|_{t=0} \Phi_t^\star \omega,$$

where Φ_t is the local flow of v on $\mathcal{M}$. Therefore $\mathfrak{L}_{\mathsf{v}}\omega = 0$ if and only if ω is constant on the integral curves of v. This is a consequence of the formula

$$\Phi_t^\star\,(\mathfrak{L}_{\mathsf{v}}\omega) = \frac{\mathsf{d}}{\mathsf{d}t}\,(\Phi_t^\star\,\omega)\,. \tag{1.23}$$

- In the particular case $k = 0$, i.e., ω is a smooth function, we get the *Lie derivative of a smooth function* $F \in \mathscr{F}(\mathcal{M})$ along a smooth vector field $\mathsf{v} \in \mathfrak{X}(\mathcal{M})$:

$$\mathfrak{L}_{\mathsf{v}}F := \mathsf{v}[F] = \mathsf{v} \lrcorner \mathsf{d}F = \mathsf{d}F(\mathsf{v}) = \left.\frac{\mathsf{d}}{\mathsf{d}t}\right|_{t=0} \Phi_t^\star F.$$

In local coordinates $x := (x_1, \dots, x_n)$ we have

$$(\mathfrak{L}_{\mathsf{v}} F)(x) := \mathsf{v}[F(x)] = \sum_{i=1}^{n} f_i(x) \frac{\partial F}{\partial x_i} = \langle f(x), \operatorname{grad}_x F(x) \rangle .$$

Here $\langle \cdot, \cdot \rangle$ is the scalar product in $\mathcal{U} \subset \mathbb{R}^n$ and

$$\operatorname{grad}_x F(x) := \left(\frac{\partial F}{\partial x_1}, \dots, \frac{\partial F}{\partial x_n} \right)^{\top}$$

is the gradient of F w.r.t. x.

- The *Lie derivative of a 2-vector field* $\mathsf{W} \in \mathfrak{X}^2(\mathcal{M})$ along $\mathsf{v} \in \mathfrak{X}(\mathcal{M})$ is the 2-vector field whose value on $F_1, F_2 \in \mathscr{F}(\mathcal{M})$ is given by

$$\mathfrak{L}_{\mathsf{v}} \mathsf{W}\,[F_1, F_2] := \mathsf{v}\,[\mathsf{W}\,[F_1, F_2]] - \mathsf{W}\,[\mathsf{v}[F_1], F_2] - \mathsf{W}\,[F_1, \mathsf{v}[F_2]] . \tag{1.24}$$

  Again we have $\mathfrak{L}_{\mathsf{v}} \mathsf{W} = 0$ if and only if W is constant on the integral curves of v.

- The *Lie derivative of a smooth vector field* $\mathsf{w} \in \mathfrak{X}(\mathcal{M})$ along $\mathsf{v} \in \mathfrak{X}(\mathcal{M})$ is defined in terms of the Lie bracket between v and w:

$$\mathfrak{L}_{\mathsf{v}} \mathsf{w} := [\mathsf{v}, \mathsf{w}].$$

  We have $\mathfrak{L}_{\mathsf{v}} \mathsf{w} = 0$ if and only if w is constant on the integral curves of v.

Example 1.14 (*On the vanishing of the Lie derivative of ω along* v)

Let us derive formula (1.23). Consider the identity

$$(\Phi_t^\star \circ \Phi_s^\star)\, \omega = \Phi_{t+s}^\star\, \omega, \qquad t, s \in \mathbb{R},$$

where $\Phi_t = \exp(t\,\mathsf{v})$, $\Phi_s = \exp(s\,\mathsf{v})$. We now take the derivative w.r.t. s at $s = 0$:

$$\left(\Phi_t^\star \circ \left.\frac{\mathsf{d}}{\mathsf{d}s}\right|_{s=0} \Phi_s^\star \right) \omega = \left.\frac{\mathsf{d}}{\mathsf{d}s}\right|_{s=0} \Phi_{t+s}^\star\, \omega,$$

namely,

$$\Phi_t^\star \mathfrak{L}_{\mathsf{v}} \omega = \frac{\mathsf{d}}{\mathsf{d}t} \Phi_t^\star\, \omega.$$

This implies that $\mathfrak{L}_{\mathsf{v}} \omega = 0$ if and only if $t \mapsto \Phi_t^\star\, \omega$ is constant, hence equal to its value at $t = 0$.

► Let $\mathcal{U} \subset \mathcal{M}$ be a coordinate chart coordinatized by $x := (x_1, \dots, x_n)$ around a point $p \in \mathcal{M}$. An *invariant function* of a flow Φ_t ($t \in \mathbb{R}$) defined on $\mathcal{U}$ is a function $F \in \mathscr{F}(\mathcal{U})$ such that $\Phi_t^\star F = F$, or equivalently,

$$F(\Phi_t(x)) = F(x) \qquad \forall\, t \in \mathbb{R},\ x \in \mathcal{U}. \tag{1.25}$$

- Condition (1.25) means that F is constant on every orbit of Φ_t. The hypersurface
$$\mathcal{S}_c := \{x \in \mathcal{U} \,:\, F(x) = c\},$$
where $c \in \mathbb{R}$ is a constant (depending on the initial condition of the initial value problem defined by v), is called *level set* of F. Therefore, if F is an invariant function under the action of Φ_t, then clearly every level set of F is an invariant set of Φ_t. However, it is not true that if the level set of a function, $\{x \in \mathcal{U} : F(x) = c\}$, is an invariant set then the function itself is an invariant function. Nevertheless it can be proved that if every level set of F is an invariant set, then F is an invariant function.

- From our previous claims we see that $F \in \mathscr{F}(\mathcal{U})$ is an invariant function of Φ_t if and only if
$$\mathsf{v}[F(x)] = (\mathfrak{L}_{\mathsf{v}} F)(x) = 0 \qquad \forall\, x \in \mathcal{U}. \tag{1.26}$$

► Let Φ_t and Ψ_s, $t, s \in \mathbb{R}$, be two distinct one-parameter Lie groups of smooth diffeomorphisms on $\mathcal{M}$ whose infinitesimal generators are respectively given by
$$\mathsf{v} := \sum_{i=1}^{n} f_i(x) \frac{\partial}{\partial x_i}, \qquad \mathsf{w} := \sum_{i=1}^{n} g_i(x) \frac{\partial}{\partial x_i}, \qquad f_i, g_i \in \mathscr{F}(\mathcal{U}),$$
with
$$f(x) := \left.\frac{\mathrm{d}}{\mathrm{d}t}\right|_{t=0} \Phi_t(x), \qquad g(x) := \left.\frac{\mathrm{d}}{\mathrm{d}s}\right|_{s=0} \Psi_s(x).$$
Then it can be proved that the flows Φ_t and Ψ_s commute if and only if
$$[\mathsf{v}, \mathsf{w}] := \sum_{i=1}^{n} \left(\mathfrak{L}_{\mathsf{v}} g_i - \mathfrak{L}_{\mathsf{w}} f_i\right)(x) \frac{\partial}{\partial x_i} = 0,$$
for all $x \in \mathcal{U}$.

► There hold the following useful formulas:
$$\mathfrak{L}_{\mathsf{v}} \mathrm{d}\omega = \mathrm{d}(\mathfrak{L}_{\mathsf{v}} \omega),$$
$$\mathfrak{L}_{\mathsf{v}} \omega = \mathrm{d}(\mathsf{v} \lrcorner\, \omega) + \mathsf{v} \lrcorner\, \mathrm{d}\omega \qquad \textit{(Cartan formula)}, \tag{1.27}$$
and
$$[\mathsf{v}_1, \mathsf{v}_2] \lrcorner\, \omega = \mathfrak{L}_{\mathsf{v}_1} (\mathsf{v}_2 \lrcorner\, \omega) - \mathsf{v}_2 \lrcorner\, \mathfrak{L}_{\mathsf{v}_1} \omega, \tag{1.28}$$
for all $\mathsf{v}, \mathsf{v}_1, \mathsf{v}_2 \in \mathfrak{X}(\mathcal{M})$ and $\omega \in \Omega^k(\mathcal{M})$. Combining (1.27) and (1.28) we also get
$$[\mathsf{v}_1, \mathsf{v}_2] \lrcorner\, \omega = \mathrm{d}(\mathsf{v}_1 \lrcorner\, \mathsf{v}_2 \lrcorner\, \omega) + \mathsf{v}_1 \lrcorner\, \mathrm{d}(\mathsf{v}_2 \lrcorner\, \omega) - \mathsf{v}_2 \lrcorner\, \mathrm{d}(\mathsf{v}_1 \lrcorner\, \omega) - \mathsf{v}_2 \lrcorner\, \mathsf{v}_1 \lrcorner\, \mathrm{d}\omega. \tag{1.29}$$

Formula (1.27) can be considered as an alternative definition of Lie derivative of a differential k-form along a smooth vector field.

Example 1.15 (*On formula (1.28)*)

Let us verify formula (1.28) when ω is a 2-form. In this case both sides of (1.28) are 1-forms. For any vector field v we have, in view of (1.22) applied to the 1-form $\mathsf{v}_2 \lrcorner \omega$,

$$\begin{aligned} \mathfrak{L}_{\mathsf{v}_1} \mathsf{v}_2 \lrcorner \omega(\mathsf{v}) &= \mathsf{v}_1[\mathsf{v}_2 \lrcorner \omega(\mathsf{v})] - \mathsf{v}_2 \lrcorner \omega([\mathsf{v}_1, \mathsf{v}]) \\ &= \mathsf{v}_1[\omega(\mathsf{v}_2, \mathsf{v})] - \omega(\mathsf{v}_2, [\mathsf{v}_1, \mathsf{v}]). \end{aligned}$$

Applying (1.22) again, but now to the 2-form ω we get

$$\begin{aligned} \mathsf{v}_2 \lrcorner \mathfrak{L}_{\mathsf{v}_1} \omega(\mathsf{v}) &= \mathfrak{L}_{\mathsf{v}_1} \omega(\mathsf{v}_2, \mathsf{v}) \\ &= \mathsf{v}_1[\omega(\mathsf{v}_2, \mathsf{v})] - \omega([\mathsf{v}_1, \mathsf{v}_2], \mathsf{v}) - \omega(\mathsf{v}_2, [\mathsf{v}_1, \mathsf{v}]). \end{aligned}$$

It follows that for any vector field v we have:

$$\mathfrak{L}_{\mathsf{v}_1} \mathsf{v}_2 \lrcorner \omega(\mathsf{v}) - \mathsf{v}_2 \lrcorner \mathfrak{L}_{\mathsf{v}_1} \omega(\mathsf{v}) = \omega([\mathsf{v}_1, \mathsf{v}_2], \mathsf{v}) = [\mathsf{v}_1, \mathsf{v}_2] \lrcorner \omega(\mathsf{v}).$$

1.3 Basic facts on Lie groups and Lie algebras

► A Lie group is a smooth manifold that is also a group such that the group operations are smooth. A Lie group is a homogeneous space in the sense that left translation by a group element h is a diffeomorphism of the group onto itself that maps the identity element to h. Therefore, locally the group looks the same around any point. To study the local structure of a Lie group, it is enough to examine a neighborhood of the identity element. It is not surprising that the tangent space at the identity of a Lie group should play a key role.

► A *group* is a set $\mathbf{G}$ together with an operation $* : \mathbf{G} \times \mathbf{G} \to \mathbf{G}$, called *group law* of $\mathbf{G}$, that combines any two elements h and k to form another element $h * k \in \mathbf{G}$. The pair $(\mathbf{G}, *)$ must satisfy four axioms:

1. *Associativity*: $(h * k) * \ell = h * (k * \ell)$ for all $h, k, \ell \in \mathbf{G}$.

2. *Existence of identity element*: there exists a unique element $e \in \mathbf{G}$, such that for every element $h \in \mathbf{G}$, the equation $h * e = e * h = h$ holds.

3. *Existence of inverse element*: for each $h \in \mathbf{G}$, there exists an element $k \in \mathbf{G}$ such that $h * k = k * h = e$.

► A *Lie group* is a smooth manifold $\mathbf{G}$ that is a group and for which the group operations of multiplication, $* : \mathbf{G} \times \mathbf{G} \to \mathbf{G} : (h, k) \mapsto h * k$, and inversion, $h \mapsto h^{-1}$, are smooth. We simplify the notation by setting $h * k \equiv h\,k$ and we always assume that $\mathbf{G}$ is finite-dimensional. If the manifold $\mathbf{G}$ has dimension n, we also say that $\mathbf{G}$ is a *n-parameter Lie group*.

- A *Lie group homomorphism* is a smooth map $\varphi : \mathbf{G} \to \mathbf{T}$ between two Lie groups which respects the group law:

$$\varphi(h\,k) = \varphi(h)\,\varphi(k), \qquad h, k \in \mathbf{G}.$$

If φ has a smooth inverse then φ is a *Lie group isomorphism*.

- It can happen that one is not interested in the full Lie group, but only in some group elements close to the identity. In this case one can dispense with the abstract manifold theory and define a local Lie group in terms of local coordinate expressions for the group operations.

    (a) A *n-parameter local Lie group* consists of connected open subsets $\mathcal{U}' \subset \mathcal{U} \subset \mathbb{R}^n$ containing the origin and smooth maps $\mathcal{U} \times \mathcal{U} \to \mathbb{R}^n$ defining the group operation, and $\mathcal{U}' \to \mathcal{U}$, defining the group inversion. The four group axioms must hold as well.

    (b) An easy method to construct local Lie groups is to take a global Lie group **G** and use a coordinate chart containing the identity element.

    (c) Less trivial is the fact that (locally) every local Lie group arises in this fashion. In other words, every local Lie group is locally isomorphic to a neighborhood of the identity of some global Lie group **G**.

- A *Lie subgroup* **T** of a Lie group **G** is an immersed submanifold of **G**, i.e., there exists a Lie group homomorphism $\varphi : \widetilde{\mathbf{T}} \to \mathbf{G}$, where $\widetilde{\mathbf{T}}$ is a Lie group, such that **T** is the image of $\widetilde{\mathbf{T}}$ under φ.

    (a) It can be proved that if **T** is a closed subgroup of the Lie group **G** then **T** is a regular submanifold of **G** and hence a Lie group. Conversely, any regular Lie subgroup of **G** is a closed subgroup.

    (b) According to this result we need only to check that **T** is a subgroup of **G** and is closed as a subset of **G** to conclude that **T** is a regular Lie subgroup. This circumvents the problem of actually proving that **T** is a submanifold.

► If **G** is a Lie group then there are certain distinguished vector fields on **G** characterized by their invariance (in a sense to be defined shortly) under the group multiplication.

- It turns out that these invariant vector fields form a n-dimensional vector space, called the *Lie algebra* $\mathfrak{g}$ of **G**.

    (a) We denote elements of $\mathfrak{g}$ by $\mathsf{X}_1, \mathsf{X}_2, \mathsf{X}_3, \ldots$ ($\mathsf{X}, \mathsf{Y}, \mathsf{Z}$ when we need to write only three elements). A basis of $\mathfrak{g}$ is denoted by $\{\mathsf{E}_1, \ldots, \mathsf{E}_n\}$.

    (b) The dual vector space of $\mathfrak{g}$, denoted by $\mathfrak{g}^*$, has elements $\mathsf{X}_1^*, \mathsf{X}_2^*, \mathsf{X}_3^*, \ldots$ ($\mathsf{X}^*, \mathsf{Y}^*, \mathsf{Z}^*$ when we need to write only three elements). A basis of $\mathfrak{g}^*$ is denoted by $\{\mathsf{E}_1^*, \ldots, \mathsf{E}_n^*\}$. With this notation we have

    $$\left\langle \mathsf{E}_i, \mathsf{E}_j^* \right\rangle = \delta_{ij}, \qquad i, j = 1, \ldots, n,$$

    but X^* is an arbitrary element of $\mathfrak{g}^*$ that is not the dual of $\mathsf{X} \in \mathfrak{g}$ (here X is an arbitrary element of $\mathfrak{g}$).

- In fact, almost all the information in the group **G** is contained in its Lie algebra $\mathfrak{g}$. This fundamental observation is the cornerstone of Lie group theory. For example, it allows us to replace complicated *nonlinear* conditions of invariance under a group action by relatively simple *linear* infinitesimal conditions.

► The construction of the Lie algebra $\mathfrak{g}$ of the Lie group **G** can be sketched as follows.

- For any group element $h \in \mathbf{G}$ the *right multiplication map*

$$R_h : \mathbf{G} \to \mathbf{G} : k \mapsto R_h(k) := k\,h$$

  is a diffeomorphism with inverse $R_{h^{-1}} = (R_h)^{-1}$.

- A vector field X on **G** is called *right-invariant* if

$$\mathsf{d}R_h(\mathsf{X}(k)) = \mathsf{X}(R_h(k)) = \mathsf{X}(k\,h),$$

  for all $h, k \in \mathbf{G}$.

- Note that if X and Y are right-invariant, so is any linear combination of X and Y. Therefore the set of right-invariant vector fields form a vector space, called the *Lie algebra* of **G**.

We constructed the Lie algebra of **G** in terms of the right multiplication map. The same can be done in terms of the left multiplication map $L_h : \mathbf{G} \to \mathbf{G} : k \mapsto L_h(k) := h\,k$. Indeed, associated with any Lie group, there are two different Lie algebras: $\mathfrak{g}_R$, spanned by right-invariant vector fields and $\mathfrak{g}_L$, spanned by left-invariant vector fields. But the following holds: if **G** is a Lie group, then the differential of the inversion map defines a Lie algebra isomorphism $\mathfrak{g}_R \simeq \mathfrak{g}_L$. The commutation between the maps L_h and R_h imply that $\mathfrak{g}_R$ (resp. $\mathfrak{g}_L$) can be characterized as the set of vector fields on **G** that commute with all left-invariant (resp. right-invariant) vector fields.

► Remarks:

- Any right-invariant vector field is uniquely determined by its value at the identity $e \in \mathbf{G}$:

$$\mathsf{X}(h) = \mathsf{d}R_h(\mathsf{X}(e)), \tag{1.30}$$

  since $R_h(e) = h$.

- Conversely, any tangent vector to **G** at e uniquely determines a right-invariant vector field on **G** by formula (1.30). Indeed we have

$$\mathsf{d}R_h(\mathsf{X}(k)) = \mathsf{d}R_h(\mathsf{d}R_k(\mathsf{X}(e))) = \mathsf{d}(R_h \circ R_k)\mathsf{X}(e) = \mathsf{d}R_{kh}\mathsf{X}(e) = \mathsf{X}(k\,h).$$

- Therefore the Lie algebra $\mathfrak{g}$ of $\mathbf{G}$ is identified with the tangent space $T_e\mathbf{G}$:
$$\mathfrak{g} \simeq T_e\mathbf{G}.$$
Similarly the cotangent space $T_e^*\mathbf{G}$ is identified with $\mathfrak{g}^*$, the dual vector space to $\mathfrak{g}$, i.e.,
$$\mathfrak{g}^* \simeq T_e^*\mathbf{G}.$$
Note that $\mathfrak{g}$ and $\mathfrak{g}^*$ are finite-dimensional vector spaces whose dimension is equal to the dimension n of $\mathbf{G}$.

- For $\mathsf{X} \in \mathfrak{g}$, $\exp(\mathsf{X})$ is the Lie group element corresponding to the group action $\Phi_1(e)$ (to be defined later), where Φ_t denotes the flow of the right-invariant vector field X on $\mathbf{G}$:
$$\Phi_1(e) = \exp(\mathsf{X})\, e.$$

- A Lie algebra $\mathfrak{g}$ is equipped with a natural skew-symmetric bilinear operation, namely the *Lie bracket*, thanks to formulas (1.6).

▶ The last observations lead to the classical definition of a Lie algebra. For consistency we define a real Lie algebra, but the definition of a complex Lie algebra can be given in a similar way.

- A finite-dimensional real *Lie algebra* $\mathfrak{g}$ is a finite-dimensional real vector space equipped with a binary mapping $[\cdot,\cdot] : \mathfrak{g} \times \mathfrak{g} \to \mathfrak{g}$, called *Lie bracket*, satisfying the following axioms:

  1. *(Bi)linearity*: $[\lambda_1 \mathsf{X} + \lambda_2 \mathsf{Y}, \mathsf{Z}] = \lambda_1[\mathsf{X},\mathsf{Z}] + \lambda_2[\mathsf{Y},\mathsf{Z}]$,
  2. *Alternating*: $[\mathsf{X},\mathsf{X}] = 0$,
  3. *Jacobi identity*: $[\mathsf{X},[\mathsf{Y},\mathsf{Z}]] + \circlearrowleft (\mathsf{X},\mathsf{Y},\mathsf{Z}) = 0$,

  for all $\mathsf{X},\mathsf{Y},\mathsf{Z} \in \mathfrak{g}$ and $\lambda_1, \lambda_2 \in \mathbb{R}$.

- Bilinearity and alternating properties imply skew-symmetry,
$$[\mathsf{X},\mathsf{Y}] = -[\mathsf{Y},\mathsf{X}],$$
while skew-symmetry only implies the alternating property. Furthermore the Jacobi identity implies that the map $\mathsf{Y} \mapsto [\mathsf{X},\mathsf{Y}]$ is a *derivation*, i.e., it satisfies the *Leibniz rule*.

- If $\{\mathsf{E}_1,\ldots,\mathsf{E}_n\}$ is a basis of $\mathfrak{g}$ then there are certain constants $c_{ij}^k \in \mathbb{R}$, $i,j,k = 1,\ldots,n$, called *structure constants* of $\mathfrak{g}$, such that
$$[\mathsf{E}_i,\mathsf{E}_j] = \sum_{k=1}^{n} c_{ij}^k\, \mathsf{E}_k. \tag{1.31}$$

Formula (1.31) clearly encodes the linear structure of $\mathfrak{g}$. The structure constants of $\mathfrak{g}$ define the Lie algebra itself. Evidently one has

$$c_{ij}^k = -c_{ji}^k,$$

for all $i, j, k = 1, \ldots, n$, and

$$\sum_{k=1}^{n} \left(c_{ij}^k \, c_{k\ell}^r + c_{\ell i}^k \, c_{kj}^r + c_{j\ell}^k \, c_{ki}^r \right) = 0,$$

for $i, j, \ell, r = 1, \ldots, n$. The last formula is nothing but the Jacobi identity for the structure constants.

Example 1.16 (*One-dimensional and two-dimensional Lie algebras*)

1. The only one-dimensional Lie algebra is Abelian. Its connected Lie groups are the line $\mathbb{R}$ and the circle $\mathbb{S}^1$.
2. There is a unique Abelian two-dimensional Lie algebra defined by the Lie bracket $[\mathsf{X}, \mathsf{Y}] = 0$, where $\{\mathsf{X}, \mathsf{Y}\}$ is a basis. This integrates, in terms of the exponential map, to three possible groups $\mathbb{R}^2, \mathbb{R} \times (\mathbb{R} \setminus \mathbb{Z})$ and $(\mathbb{R} \setminus \mathbb{Z})^2$.

► A *subalgebra* $\mathfrak{t}$ of a Lie algebra $\mathfrak{g}$ is a vector subspace which is closed under the Lie bracket, so that $[\mathsf{X}, \mathsf{Y}] \in \mathfrak{t}$ whenever $\mathsf{X}, \mathsf{Y} \in \mathfrak{t}$.

- If $\mathbf{T}$ is a Lie subgroup of a Lie group $\mathbf{G}$, any right-invariant vector field X on $\mathbf{T}$ can be extended to a right-invariant vector field on $\mathbf{G}$ setting $\mathsf{X}(h) = \mathsf{d}R_h(\mathsf{X}(e))$, $h \in \mathbf{G}$. In this way the Lie subalgebra $\mathfrak{t}$ is realized as subalgebra of the Lie algebra $\mathfrak{g}$ of $\mathbf{G}$.
- It can be proved that if $\mathbf{T}$ is a Lie subgroup of $\mathbf{G}$, its Lie algebra $\mathfrak{t}$ is a Lie subalgebra of $\mathfrak{g}$. Conversely, if $\mathfrak{t}$ is any subalgebra of $\mathfrak{g}$, then there is a unique connected Lie subgroup $\mathbf{T}$ of $\mathbf{G}$ with Lie algebra $\mathfrak{t}$.

1.3.1 Matrix Lie groups and matrix Lie algebras

► The main examples of Lie groups include linear groups (*matrix Lie groups*), i.e., Lie subgroups of the *general linear Lie group*

$$\mathbf{GL}(n, \mathbb{R}) := \{h \in \mathsf{End}(\mathbb{R}^n) \ : \ \det h \neq 0\},$$

the (non-connected) non-commutative group of all invertible $n \times n$ matrices (with coefficients in $\mathbb{R}$), where the group operation is given by product of matrices.

- $\mathbf{GL}(n, \mathbb{R})$ is a real smooth manifold of dimension n^2 since it is an open subset of the vector space $\mathsf{End}(\mathbb{R}^n)$ of all linear maps (not necessarily invertible) from $\mathbb{R}^n$ to $\mathbb{R}^n$. Similarly one can define $\mathbf{GL}(n, \mathbb{C})$.

- One can define $\mathbf{GL}(n,\mathbb{R})$ as the inverse image of $\mathbb{R}\setminus\{0\}$ under the submersion $h \mapsto \det h$. This is true thanks to the "Regular value Theorem".

► The main examples of Lie algebras include *matrix Lie algebras*, i.e., Lie subalgebras of *general linear Lie algebra* $\mathfrak{gl}(n,\mathbb{R}) \simeq \mathrm{End}(\mathbb{R}^n)$, the vector space of all linear maps (not necessarily invertible) from $\mathbb{R}^n$ to $\mathbb{R}^n$. In $\mathfrak{gl}(n,\mathbb{R})$ the Lie bracket is given by the commutator of matrices.

- $\mathfrak{gl}(n,\mathbb{R})$ is the Lie algebra of $\mathbf{GL}(n,\mathbb{R})$. Such a connection is established in terms of the *exponential map*, which is, in this case, the usual exponential of matrices,

$$\exp(\mathsf{X}) = \sum_{k=0}^{\infty} \frac{\mathsf{X}^k}{k!} = 1_n + \mathsf{X} + \frac{1}{2!}\mathsf{X}^2 + \frac{1}{3!}\mathsf{X}^3 + \dots, \qquad \mathsf{X} \in \mathfrak{gl}(n,\mathbb{R}).$$

  Here 1_n is the identity element in $\mathbf{GL}(n,\mathbb{R})$, i.e., the $n \times n$ identity matrix.

- According to "Ado Theorem", every finite-dimensional Lie algebra is isomorphic to a matrix Lie algebra (i.e., for every finite-dimensional matrix Lie algebra there is a matrix Lie group with this algebra as its Lie algebra). The corresponding theorem does not hold true for (finite-dimensional) Lie groups: not every Lie group is isomorphic to a subgroup of $\mathbf{GL}(n,\mathbb{R})$.

- A consequence of "Ado Theorem" establishes the fundamental correspondence between Lie groups and Lie algebras: if $\mathfrak{g}$ is a finite-dimensional Lie algebra then there exists a unique (connected) Lie group $\mathbf{G}$ having $\mathfrak{g}$ as its Lie algebra.

Example 1.17 (*The orthogonal Lie group and its Lie algebra*)

The *orthogonal Lie group*

$$\mathbf{O}(n) := \left\{ h \in \mathbf{GL}(n,\mathbb{R}) \,:\, h h^\top = h^\top h = 1_n \right\}$$

is a smooth manifold of dimension $n(n-1)/2$. It is a non-connected subgroup of $\mathbf{GL}(n,\mathbb{R})$. One has $\det h = \pm 1$, $h \in \mathbf{O}(n)$. A connected subgroup of $\mathbf{O}(n)$ is the *special orthogonal Lie group*:

$$\mathbf{SO}(n) := \{ h \in \mathbf{O}(n) \,:\, \det h = 1 \}.$$

- Let us construct the matrix Lie algebra of $\mathbf{O}(n)$, called the *orthogonal Lie algebra*. For $\mathsf{X} \in \mathfrak{gl}(n,\mathbb{R})$ and for small $|t|$, consider the invertible matrix

$$\exp(t\mathsf{X}) = 1_n + t\mathsf{X} + O(t^2).$$

  Orthogonality of $\exp(t\mathsf{X})$ yields

$$\begin{aligned} 1_n &= \exp(t\mathsf{X})(\exp(t\mathsf{X}))^\top \\ &= \left(1_n + t\mathsf{X} + O(t^2)\right)\left(1_n + t\mathsf{X}^\top + O(t^2)\right) \\ &= 1_n + t\left(\mathsf{X} + \mathsf{X}^\top\right) + O(t^2), \end{aligned}$$

which gives

$$T_{1_n}\mathbf{O}(n) = \mathfrak{o}(n) = \left\{\mathsf{X} \in \mathfrak{gl}(n,\mathbb{R}) \,:\, \mathsf{X} + \mathsf{X}^\top = 0_n\right\}.$$

It turns out that $T_{1_n}\mathbf{SO}(n) = \mathfrak{so}(n) \simeq \mathfrak{o}(n)$.

- Fix $n = 3$. A possible choice of a basis in $\mathfrak{o}(3)$ is

$$\mathsf{E}_1 := \begin{pmatrix} 0 & 0 & 0 \\ 0 & 0 & -1 \\ 0 & 1 & 0 \end{pmatrix}, \quad \mathsf{E}_2 := \begin{pmatrix} 0 & 0 & 1 \\ 0 & 0 & 0 \\ -1 & 0 & 0 \end{pmatrix}, \quad \mathsf{E}_3 := \begin{pmatrix} 0 & -1 & 0 \\ 1 & 0 & 0 \\ 0 & 0 & 0 \end{pmatrix}.$$

Note that

$$[\,\mathsf{E}_i, \mathsf{E}_j\,] = \varepsilon_{ijk}\mathsf{E}_k, \tag{1.32}$$

where ε_{ijk} is a totally skew-symmetric tensor with $\varepsilon_{123} = 1$. Here ε_{ijk} defines the structure constants of $\mathfrak{o}(3)$.

- We can explicitly describe the corresponding subgroups in $\mathbf{O}(3)$. For instance,

$$\exp(t\,\mathsf{E}_1) = \begin{pmatrix} 1 & 0 & 0 \\ 0 & \cos t & -\sin t \\ 0 & \sin t & \cos t \end{pmatrix},$$

which is a rotation around the axis 1 by angle t. Similarly, E_2 and E_3 generate rotations around axes 2 and 3.

Example 1.18 (*The special linear Lie group and its Lie algebra*)

The real *special linear Lie group*

$$\mathbf{SL}(n,\mathbb{R}) := \{h \in \mathbf{GL}(n,\mathbb{R}) \,:\, \det h = 1\} = \det{}^{-1}(1)$$

is a smooth manifold of dimension $n^2 - 1$. It is a connected subgroup of $\mathbf{GL}(n,\mathbb{R})$.

- Let $h, k \in \mathbf{GL}(n,\mathbb{R})$. It can be proved that the differential of the submersion $h \mapsto \det h$ is

$$(\mathrm{d}\det h)(k) = \det h\,\mathrm{Trace}\left(h^{-1}k\right).$$

- The Lie algebra of $\mathbf{SL}(n,\mathbb{R})$ is the (n^2-1)-dimensional vector space

$$T_{1_n}\mathbf{SL}(n,\mathbb{R}) = \mathfrak{sl}(n,\mathbb{R}) = \mathrm{Ker}(\mathrm{d}\det h) = \{\mathsf{X} \in \mathfrak{gl}(n,\mathbb{R}) \,:\, \mathrm{Trace}\,\mathsf{X} = 0\}.$$

This is called *special linear Lie algebra*.

1.3.2 Lie group actions

► Lie groups appear most often through their action (taken by us as the left action) on a smooth manifold $\mathcal{M}$.

► Let $\mathcal{M}$ be a n-dimensional smooth manifold and $\mathbf{G}$ be a r-dimensional Lie group. A diffeomorphism

$$\varphi : \mathbf{G} \times \mathcal{M} \to \mathcal{M}$$

is an *action of* $\mathbf{G}$ *on* $\mathcal{M}$ if:

1. $\varphi(e, p) = p$ for all $p \in \mathcal{M}$;

2. $\varphi(h, \varphi(k,p)) = \varphi(hk, p)$ for all $h, k \in \mathbf{G}$ and for all $p \in \mathcal{M}$.

Example 1.19 (*Lie group actions*)

1. The Lie group $\mathbf{GL}(3,\mathbb{R})$ acts on $\mathbb{R}^3$ by $(h,x) \mapsto h\,x$, $h \in \mathbf{GL}(3,\mathbb{R})$, $x \in \mathbb{R}^3$.
2. A global flow $\Phi_t : \mathbb{R} \times \mathcal{M} \to \mathcal{M}$ generated by a compactly supported vector field is an action of $\mathbb{R}$ on $\mathcal{M}$. It defines a global Lie group of transformations (i.e., a globally defined continuous dynamical system).
3. A local flow $\Phi_t : (-\varepsilon, \varepsilon) \times \mathcal{M} \to \mathcal{M}$ generated by a vector field is a local action of $(-\varepsilon, \varepsilon)$ on $\mathcal{M}$. It defines a local Lie group of transformations (i.e., a locally defined continuous dynamical system).

- The triple $(\varphi, \mathbf{G}, \mathcal{M})$ is called *Lie group of transformations* and *local Lie group of transformations* if $\mathbf{G}$ is a local Lie group.
- A set $\mathcal{S} \subset \mathcal{M}$ is a $\mathbf{G}$*-invariant set* if $\varphi(h,p) \in \mathcal{S}$ for all $h \in \mathbf{G}$ and $p \in \mathcal{S}$. A smooth function $F \in \mathscr{F}(\mathcal{M})$ is a $\mathbf{G}$*-invariant function* if

$$F(\varphi(h,p)) = F(p),$$

for all $h \in \mathbf{G}$ and $p \in \mathcal{M}$. It can be proved that $F \in \mathscr{F}(\mathcal{M})$ is $\mathbf{G}$-invariant if and only if every level set $\{p \in \mathcal{M} \,:\, F(p) = c\}$, with c constant, is $\mathbf{G}$-invariant.

- The action allows us to associate with each element $\mathsf{X} \in \mathfrak{g}$ a vector field $\underline{\mathsf{X}}$ on $\mathcal{M}$, whose value at $p \in \mathcal{M}$, is the derivation on $\mathscr{F}(\mathcal{M})$ at p given by

$$\underline{\mathsf{X}}(p)[F] := \left.\frac{\mathsf{d}}{\mathsf{d}t}\right|_{t=0} F\left(\varphi(\exp(t\mathsf{X}), p)\right) \qquad \forall\, F \in \mathscr{F}(\mathcal{M}). \tag{1.33}$$

The vector field $\underline{\mathsf{X}} \in \mathfrak{X}(\mathcal{M})$ is called *fundamental vector field* (or *infinitesimal generator of* $\mathbf{G}$) corresponding to $\mathsf{X} \in \mathfrak{g}$. Its flow is given by the action of the one-parameter group $\exp(t\mathsf{X})$. We will give conditions on the action φ in such a way that we are allowed to not distinguish between the elements of the Lie algebra $\mathfrak{g}$ and the infinitesimal generators of $\mathbf{G}$.

- The construction of fundamental vector fields can be naturally generalized to *fundamental multi-vector fields* on $\mathcal{M}$.

  (a) Introduce the *exterior algebra* of $\mathfrak{g}$:

$$\mathscr{E}(\mathfrak{g}) := \bigoplus_{k \geqslant 0} \mathfrak{g}^{\wedge k}. \tag{1.34}$$

  An element of $\mathfrak{g}^{\wedge k}$ is a multi-vector of the form $\mathsf{X} = \mathsf{X}_1 \wedge \cdots \wedge \mathsf{X}_k$, $\mathsf{X}_i \in \mathfrak{g}$. One has

$$\mathsf{X} \wedge \mathsf{Y} = (-1)^{k\ell}\mathsf{Y} \wedge \mathsf{X}, \qquad \mathsf{X} \in \mathfrak{g}^{\wedge k},\ \mathsf{Y} \in \mathfrak{g}^{\wedge \ell}.$$

(b) The action φ allows us also to associate to each element $\mathsf{X} \in \mathfrak{g}^{\wedge k}$ a multivector field $\underline{\mathsf{X}} \in \mathfrak{X}^k(\mathcal{M})$ on $\mathcal{M}$, defined by $\underline{\mathsf{X}} := \underline{\mathsf{X}}_1 \wedge \cdots \wedge \underline{\mathsf{X}}_k$, $\underline{\mathsf{X}}_i \in \mathfrak{X}(\mathcal{M})$.

► We give some important definitions:

- The *orbit* through $p \in \mathcal{M}$ of $(\varphi, \mathbf{G}, \mathcal{M})$ is:

$$\mathcal{O}_p := \{\varphi(h,p) \,:\, h \in \mathbf{G}\}.$$

  This is the minimal non-empty $\mathbf{G}$-invariant subset of $\mathcal{M}$ containing p.

- The *isotropy group* (or *stabilizer*) of $p \in \mathcal{M}$ is:

$$\mathbf{G}_p := \{h \in \mathbf{G} \,:\, \varphi(h,p) = p\}.$$

  This is the subgroup of $\mathbf{G}$ consisting of all elements $h \in \mathbf{G}$ such that the action φ fixes $p \in \mathcal{M}$. The *global isotropy subgroup* is:

$$\mathbf{G}_{\mathcal{M}} := \bigcap_{p \in \mathcal{M}} \mathbf{G}_p = \{h \in \mathbf{G} \,:\, \varphi(h,p) = p,\ p \in \mathcal{M}\}.$$

- φ is *effective* (or *faithful*) if the equality $\varphi(h,p) = \varphi(k,p)$ for any $p \in \mathcal{M}$ holds if and only if $h = k$. This is equivalent to the statement that the only group element acting as the identity transformation is e. The effectiveness of φ is measured by $\mathbf{G}_{\mathcal{M}}$: φ is effective if and only if $\mathbf{G}_{\mathcal{M}} = \{e\}$. More generally, φ is *locally effective* if $\mathbf{G}_{\mathcal{M}}$ is a discrete subgroup of $\mathbf{G}$, which is equivalent to the existence of a neighborhood $\mathcal{U}$ of the identity e such that $\mathbf{G}_{\mathcal{M}} \cap \mathcal{U} = \{e\}$.

- φ is *free* if all isotropy groups are trivial, i.e., $\mathbf{G}_p = \{e\}$ for all $p \in \mathcal{M}$. φ is *locally free* if this holds for all $h \neq e$ in a neighborhood of e. Note that a free action is effective, the converse is not true.

- φ is *transitive* if it has only one orbit.

- A *homogeneous space* is a manifold with a transitive action φ that acts on it.

► Important facts about Lie group actions are:

- Even if $\mathbf{G}$ does not act effectively, we can, without any significant loss of information or generality, replace it by the quotient group $\mathbf{G}/\mathbf{G}_{\mathcal{M}}$. Then $\mathbf{G}/\mathbf{G}_{\mathcal{M}}$ does act effectively. For a locally effective action, the quotient group $\mathbf{G}/\mathbf{G}_{\mathcal{M}}$ is a Lie group having the same dimension, and the same local structure as $\mathbf{G}$ itself.

- $\mathbf{G}$ acts effectively freely if and only if $\mathbf{G}/\mathbf{G}_{\mathcal{M}}$ acts freely. This is equivalent to saying that $\mathbf{G}_p = \mathbf{G}_{\mathcal{M}}$ for all $p \in \mathcal{M}$.

- An action $\varphi : \mathbf{G} \times \mathcal{M} \to \mathcal{M}$ is global and transitive if and only if $\mathcal{M}$ is isomorphic to the homogeneous space $\mathbf{G}/\mathbf{G}_p$, where $p \in \mathcal{M}$.

- An r-dimensional Lie group $\mathbf{G}$ acts locally free if and only if its orbits have dimension r. It acts effectively freely if and only if its orbits have dimension $r - \dim \mathbf{G}_{\mathcal{M}}$.

► The fundamental vector fields describe infinitesimally the action of $\mathbf{G}$ on $\mathcal{M}$ and they span the tangent space to the orbits of $\mathbf{G}$ at every point of $\mathcal{M}$. This explains why the fundamental vector fields are also called infinitesimal generators of $\mathbf{G}$.

- In general, the infinitesimal generators of the group action are not isomorphic to the Lie algebra $\mathfrak{g}$ since some of the nonzero Lie algebra elements may map to the trivial (zero) vector field on $\mathcal{M}$.

- Nevertheless if the action is locally effective one can indeed identify $\underline{\mathsf{X}}$ with X. The linear map (1.33), say τ, mapping an element $\mathsf{X} \in \mathfrak{g}$ to the corresponding vector field $\underline{\mathsf{X}} = \tau(\mathsf{X})$ on $\mathcal{M}$ defines a Lie algebra homomorphism:

  $$\tau([\mathsf{X}, \mathsf{Y}]) = [\tau(\mathsf{X}), \tau(\mathsf{Y})] \qquad \forall \mathsf{X}, \mathsf{Y} \in \mathfrak{g}.$$

  Moreover, the image of $\mathfrak{g}$ under τ is a finite-dimensional Lie algebra of vector fields on $\mathcal{M}$ that is isomorphic to the Lie algebra of the effectively acting quotient group $\mathbf{G}/\mathbf{G}_{\mathcal{M}}$.

- Thus, if $\{\mathsf{E}_1, \ldots, \mathsf{E}_r\}$ forms a basis for $\mathfrak{g}$, the condition for local effectiveness is that the corresponding vector fields $\underline{\mathsf{E}}_k = \tau(\mathsf{E}_k)$ be linearly independent over $\mathbb{R}$.

- The following result holds: let $\mathfrak{g}$ be a finite-dimensional Lie algebra of vector fields on a manifold $\mathcal{M}$. Let $\mathbf{G}$ denote a Lie group having Lie algebra $\mathfrak{g}$. Then there is a local action of $\mathbf{G}$ whose infinitesimal generators coincide with the given Lie algebra.

1.3.3 Representations

► The simplest action of a Lie group $\mathbf{G}$ on a vector space $\mathcal{V}$ (not necessarily of finite dimension) is a *linear action*, which means that for each $h \in \mathbf{G}$ one has that $\varphi_h \in \mathbf{GL}(\mathcal{V})$. One can view φ as a homomorphism $\varphi : \mathbf{G} \to \mathbf{GL}(\mathcal{V})$ and one says that φ is a *representation of* $\mathbf{G}$ on $\mathcal{V}$.

► The same terminology applies to Lie algebra representations.

- A *representation of* $\mathfrak{g}$ on $\mathcal{V}$ is a Lie algebra homomorphism $\mathfrak{g} \to \mathsf{End}(\mathcal{V})$, the Lie bracket in $\mathsf{End}(\mathcal{V})$ being the commutator of endomorphisms. In other words, a linear map $\varphi : \mathfrak{g} \to \mathsf{End}(\mathcal{V})$ is a representation of $\mathfrak{g}$ on $\mathcal{V}$ if

$$\varphi([\mathsf{X},\mathsf{Y}]) = [\varphi(\mathsf{X}), \varphi(\mathsf{Y})]$$

for all $\mathsf{X},\mathsf{Y} \in \mathfrak{g}$. In such a case one says that $\mathcal{V}$ is a $\mathfrak{g}$-*module*. For $\mathsf{v} \in \mathcal{V}$ one writes $\varphi(\mathsf{X})(\mathsf{v})$ as $\mathsf{X.v}$.

- Using the fact that $\mathsf{End}(\mathcal{V})$ is the Lie algebra of $\mathbf{GL}(\mathcal{V})$ and using our convention that we identify all tangent spaces to $\mathbf{G}$ with $\mathfrak{g}$, every representation of $\mathbf{G}$ on $\mathcal{V}$ leads to a representation of $\mathfrak{g}$ on $\mathcal{V}$ by mapping $\mathsf{X} \in \mathfrak{g}$ to $\underline{\mathsf{X}} \in \mathsf{End}(\mathcal{V})$.

► We now provide some useful notions and facts coming from cohomology theory of Lie algebras (note the analogies with the differential d and with the de Rham cohomology theory).

- Let $\varphi : \mathfrak{g} \to \mathsf{End}(\mathcal{V})$ be a representation of $\mathfrak{g}$ on $\mathcal{V}$. For $k \in \mathbb{N}$ we define a k-*cochain* of $\mathfrak{g}$ as a k-linear skew-symmetric map from $\mathfrak{g}$ to $\mathcal{V}$. A 0-cochain is just an element of $\mathcal{V}$, a 1-cochain is a linear map from $\mathfrak{g}$ to $\mathcal{V}$.

- We introduce the space of k-cochains of $\mathfrak{g}$:

$$\mathcal{C}^k(\mathfrak{g},\mathcal{V}) := \mathsf{Hom}\left(\mathfrak{g}^{\wedge k},\mathcal{V}\right).$$

Note that $\mathcal{C}^0(\mathfrak{g},\mathcal{V}) \simeq \mathcal{V}$.

- Let $r \in \mathcal{C}^k(\mathfrak{g},\mathcal{V})$. The *coboundary operator* $\delta : \mathcal{C}^k(\mathfrak{g},\mathcal{V}) \to \mathcal{C}^{k+1}(\mathfrak{g},\mathcal{V})$ is defined by the map (cf. (1.17))

$$\begin{aligned} \delta r(\mathsf{X}_0,\ldots,\mathsf{X}_k) \quad &:= \quad \sum_{i=0}^{k}(-1)^i \mathsf{X}_i.\left(r\left(\mathsf{X}_0,\ldots,\not{\mathsf{X}_i},\ldots,\mathsf{X}_k\right)\right) \\ &\quad + \sum_{i<j}(-1)^{i+j} r\left([\mathsf{X}_i,\mathsf{X}_j],\mathsf{X}_0,\ldots,\not{\mathsf{X}_i},\ldots,\not{\mathsf{X}_j},\ldots,\mathsf{X}_k\right), \end{aligned} \tag{1.35}$$

where $\mathsf{X}_i \in \mathfrak{g}$, $i = 0,\ldots,k$. There holds $\delta^2 = \delta \circ \delta = 0$.

- Let $r \in \mathcal{C}^k(\mathfrak{g},\mathcal{V})$.

  (a) r is called a k-*cocycle* if $\delta r = 0$.

  (b) r is called k-*coboundary* if there exists $s \in \mathcal{C}^{k-1}(\mathfrak{g},\mathcal{V})$ such that $\delta s = r$.

Example 1.20 (*Coboundary operators for k = 0,1*)

Consider formula (1.35).

- For $k = 0$ one has
$$\delta r(\mathsf{X}) := \mathsf{X}.r.$$
- For $k = 1$ one has
$$\delta r(\mathsf{X}_1, \mathsf{X}_2) := \mathsf{X}_1.r(\mathsf{X}_2) - \mathsf{X}_2.r(\mathsf{X}_1) - r([\mathsf{X}_1, \mathsf{X}_2]).$$

1.3.4 Adjoint actions

► Two important examples of actions are the *adjoint* (resp. *coadjoint action*) of a Lie group $\mathbf{G}$ (and of a Lie algebra $\mathfrak{g}$) on its Lie algebra $\mathfrak{g}$ (resp. on $\mathfrak{g}^*$).

- For $h \in \mathbf{G}$ we define Ad_h to be the endomorphism of $\mathfrak{g}$ that is the derivative of the *conjugation map*
$$C_h := L_h \circ R_h : \mathbf{G} \to \mathbf{G} : k \mapsto hkh^{-1}$$
at the identity, i.e., $\mathsf{Ad}_h := \mathsf{d}C_h(e)$. The *adjoint action of* $\mathbf{G}$ *on* $\mathfrak{g}$ is then given by
$$\mathsf{Ad} : \mathbf{G} \to \mathbf{GL}(\mathfrak{g}) : h \mapsto \mathsf{Ad}_h := \mathsf{d}C_h(e).$$
This is a representation of $\mathbf{G}$ on $\mathfrak{g}$. If $\mathbf{G}$ is a matrix Lie group and $\mathfrak{g}$ its matrix Lie algebra one has
$$\mathsf{Ad}_h \mathsf{X} = \mathsf{d}C_h(e)(\mathsf{X}) = h\mathsf{X}h^{-1}, \qquad h \in \mathbf{G}, \mathsf{X} \in \mathfrak{g}.$$
- The representation of $\mathfrak{g}$ on itself that corresponds to the adjoint action is called the *adjoint representation of* $\mathfrak{g}$ *on itself* and is denoted by
$$\mathsf{ad} : \mathfrak{g} \to \mathsf{End}(\mathfrak{g}).$$
The image of $\mathsf{X} \in \mathfrak{g}$ under ad is denoted by ad_X. By construction $\mathsf{ad}_\mathsf{X} : \mathfrak{g} \to \mathfrak{g}$ is the fundamental vector field $\underline{\mathsf{X}}$ on $\mathfrak{g}$ that corresponds to the adjoint action (see (1.33)), viewed as an endomorphism of $\mathfrak{g}$ (by identifying all tangent spaces of $\mathfrak{g}$ to $\mathfrak{g}$):
$$\mathsf{ad}_\mathsf{X} \mathsf{Y} := \left.\frac{\mathsf{d}}{\mathsf{d}t}\right|_{t=0} \left(\mathsf{Ad}_{\exp(t\mathsf{X})} \mathsf{Y}\right) = [\mathsf{X}, \mathsf{Y}], \qquad \mathsf{X}, \mathsf{Y} \in \mathfrak{g}. \tag{1.36}$$
See Example 1.21 for a derivation of the second equality in the case of matrix Lie groups.

  (a) One can prove that
$$\mathsf{Ad}_{\exp(\mathsf{X})} = \exp\left(\mathsf{ad}_\mathsf{X}\right), \qquad \mathsf{X} \in \mathfrak{g}. \tag{1.37}$$

(b) The Jacobi identity on $\mathfrak{g}$ implies

$$\mathsf{ad}_{[\mathsf{Y},\mathsf{X}]} = [\mathsf{ad}_\mathsf{Y}, \mathsf{ad}_\mathsf{X}] \qquad \forall \mathsf{X},\mathsf{Y} \in \mathfrak{g}. \tag{1.38}$$

Indeed, for $\mathsf{Z} \in \mathfrak{g}$,

$$\mathsf{ad}_{[\mathsf{Y},\mathsf{X}]}\mathsf{Z} = [[\mathsf{Y},\mathsf{X}],\mathsf{Z}],$$

and

$$[\mathsf{ad}_\mathsf{Y}, \mathsf{ad}_\mathsf{X}]\mathsf{Z} = \mathsf{ad}_\mathsf{Y}[\mathsf{X},\mathsf{Z}] - \mathsf{ad}_\mathsf{X}[\mathsf{Y},\mathsf{Z}] = [\mathsf{Y},[\mathsf{X},\mathsf{Z}]] - [\mathsf{X},[\mathsf{Y},\mathsf{Z}]].$$

- For $h \in \mathbf{G}$, we define the map Ad^*_h by

$$\langle \mathsf{Ad}^*_h \mathsf{X}^*, \mathsf{X} \rangle := \langle \mathsf{X}^*, \mathsf{Ad}_{h^{-1}} \mathsf{X} \rangle, \qquad \mathsf{X} \in \mathfrak{g}, \mathsf{X}^* \in \mathfrak{g}^*. \tag{1.39}$$

The map

$$\mathsf{Ad}^* : \mathbf{G} \to \mathbf{GL}(\mathfrak{g}^*) : h \mapsto \mathsf{Ad}^*_h$$

is called the *coadjoint action of* $\mathbf{G}$ *on* $\mathfrak{g}^*$. Its orbits are called *coadjoint orbits*. If $\mathsf{X}^* \in \mathfrak{g}^*$ the coadjoint orbit through X^* is

$$\mathcal{O}_{\mathsf{X}^*} := \{\mathsf{Ad}^*_h \mathsf{X}^* \ : \ h \in \mathbf{G}\}.$$

- Finally, the representation of $\mathfrak{g}$ on $\mathfrak{g}^*$, denoted by ad^*, is the *coadjoint representation of* $\mathfrak{g}$ *on* $\mathfrak{g}^*$:

$$\mathsf{ad}^* : \mathfrak{g} \to \mathsf{End}(\mathfrak{g}^*).$$

By definition we have

$$\mathsf{ad}^*_\mathsf{X} \mathsf{X}^* := \left.\frac{\mathsf{d}}{\mathsf{d}t}\right|_{t=0} \left(\mathsf{Ad}^*_{\exp(t\mathsf{X})} \mathsf{X}^*\right), \qquad \mathsf{X} \in \mathfrak{g}, \mathsf{X}^* \in \mathfrak{g}^*.$$

There follows that ad and ad^* are related by the following formulas

$$\langle \mathsf{ad}^*_\mathsf{X} \mathsf{X}^*, \mathsf{Y} \rangle = -\langle \mathsf{X}^*, \mathsf{ad}_\mathsf{X} \mathsf{Y} \rangle = \langle \mathsf{X}^*, [\mathsf{Y},\mathsf{X}] \rangle, \qquad \mathsf{X}^* \in \mathfrak{g}^*, \mathsf{X},\mathsf{Y} \in \mathfrak{g}. \tag{1.40}$$

Note that

$$\langle \mathsf{ad}^*_\mathsf{X} \mathsf{X}^*, \mathsf{Y} \rangle = -\langle \mathsf{ad}^*_\mathsf{Y} \mathsf{X}^*, \mathsf{X} \rangle. \tag{1.41}$$

Example 1.21 (*Derivation of (1.36) for matrix Lie groups*)

Let us derive formula (1.36) in the case matrix Lie groups. By definition we have:

$$\mathsf{ad}_\mathsf{X}\mathsf{Y} := \left.\frac{\mathsf{d}}{\mathsf{d}t}\right|_{t=0} \left(\mathsf{Ad}_{\exp(t\mathsf{X})} \mathsf{Y}\right), \qquad \mathsf{X},\mathsf{Y} \in \mathfrak{g}. \tag{1.42}$$

Formula (1.42) has to be compared with (1.33) where $\varphi = \mathsf{Ad}$.

- Let $h(t) := \exp(t\mathsf{X})$, $\mathsf{X} \in \mathfrak{gl}(n,\mathbb{R})$. Then $h(0) = 1_n$ and $(\mathsf{d}/\mathsf{d}t)|_{t=0} h(t) = \mathsf{X}$. Thus if $\mathsf{Y} \in$

$\mathfrak{gl}(n,\mathbb{R})$, we have

$$\begin{aligned}\left.\frac{\mathsf{d}}{\mathsf{d}t}\right|_{t=0}\left(\mathsf{Ad}_{\exp(t\mathsf{X})}\,\mathsf{Y}\right) &= \left.\frac{\mathsf{d}}{\mathsf{d}t}\right|_{t=0}\left(\exp(t\mathsf{X})\,\mathsf{Y}\,(\exp(t\mathsf{X}))^{-1}\right)\\ &= \left.\frac{\mathsf{d}}{\mathsf{d}t}\right|_{t=0}\left(h(t)\,\mathsf{Y}\,h^{-1}(t)\right)\\ &= \left(\left.\frac{\mathsf{d}}{\mathsf{d}t}\right|_{t=0} h(t)\right)\mathsf{Y}\,h^{-1}(0) + h(0)\,\mathsf{Y}\left(\left.\frac{\mathsf{d}}{\mathsf{d}t}\right|_{t=0} h^{-1}(t)\right).\end{aligned}$$

- Differentiating the identity $h(t)\,h^{-1}(t) = 1_n$ we get

$$\left(\frac{\mathsf{d}}{\mathsf{d}t}h(t)\right)h^{-1}(t) + h(t)\left(\frac{\mathsf{d}}{\mathsf{d}t}h^{-1}(t)\right) = 0_n,$$

so that

$$\frac{\mathsf{d}}{\mathsf{d}t}h^{-1}(t) = -h^{-1}(t)\left(\frac{\mathsf{d}}{\mathsf{d}t}h(t)\right)h^{-1}(t).$$

- Hence we have

$$\left.\frac{\mathsf{d}}{\mathsf{d}t}\right|_{t=0} h^{-1}(t) = -\left.\frac{\mathsf{d}}{\mathsf{d}t}\right|_{t=0} h(t) = -\mathsf{X}.$$

- Therefore we find

$$\left.\frac{\mathsf{d}}{\mathsf{d}t}\right|_{t=0}\left(\mathsf{Ad}_{\exp(t\mathsf{X})}\,\mathsf{Y}\right) = \mathsf{X}\mathsf{Y} - \mathsf{Y}\mathsf{X} = \mathsf{ad}_{\mathsf{X}}\mathsf{Y}.$$

► Let us define the so-called *tensor algebra* of $\mathfrak{g}$:

$$\mathscr{T}(\mathfrak{g}) := \bigoplus_{k\geqslant 0}\mathfrak{g}^{\otimes k}, \qquad \mathfrak{g}^{\otimes k} := \underbrace{\mathfrak{g}\otimes\cdots\otimes\mathfrak{g}}_{k\text{ times}}. \tag{1.43}$$

- The multiplication in $\mathscr{T}(\mathfrak{g})$ is determined by the canonical isomorphism $\mathfrak{g}^{\otimes k}\otimes\mathfrak{g}^{\otimes \ell}\to\mathfrak{g}^{\otimes(k+\ell)}$ given by the tensor product, which is then extended by linearity to all of $\mathscr{T}(\mathfrak{g})$. This multiplication rule implies that $\mathscr{T}(\mathfrak{g})$ is naturally a graded algebra with $\mathfrak{g}^{\otimes k}$ serving as the k-th grade subspace.
- Many other algebras of interest can be constructed starting with the tensor algebra and then imposing certain relations on the generators, i.e., by constructing certain quotient algebras of $\mathscr{T}(\mathfrak{g})$. Examples of this are the exterior algebra already defined in (1.34) and the symmetric algebra:

  (a) The *exterior algebra* $\mathscr{E}(\mathfrak{g})$ defined in (1.34) is the quotient algebra of $\mathscr{T}(\mathfrak{g})$ by the two-sided ideal I generated by all elements of the form $\mathsf{X}\otimes\mathsf{X}$ with $\mathsf{X}\in\mathfrak{g}$. The exterior product $\wedge$ of two elements of $\mathscr{E}(\mathfrak{g})$ is defined by $\mathsf{X}\wedge\mathsf{Y} := \mathsf{X}\otimes\mathsf{Y} \pmod{I}$, $\mathsf{X},\mathsf{Y}\in\mathfrak{g}$, where $\pmod{I}$ means that we do the tensor product in the usual way and then declare every element of the tensor that is in the ideal to be zero.

  (b) The *symmetric algebra* $\mathscr{S}(\mathfrak{g})$ is the quotient algebra of $\mathscr{T}(\mathfrak{g})$ by the two-sided ideal generated by all elements of the form $\mathsf{X}\otimes\mathsf{Y}-\mathsf{Y}\otimes\mathsf{X}$ with $\mathsf{X},\mathsf{Y}\in\mathfrak{g}$.

► Adjoint actions can be generalized to $\mathscr{T}(\mathfrak{g})$. Let us briefly explain this generalization.

- Let $\mathsf{X}_1 \otimes \cdots \otimes \mathsf{X}_k \in \mathfrak{g}^{\otimes k}$, $\mathsf{X}_i \in \mathfrak{g}$. Then we generalize the map $\mathsf{ad} : \mathfrak{g} \to \mathsf{End}(\mathfrak{g})$ to a map
$$\mathsf{ad}^{(k)} : \mathfrak{g} \to \mathsf{End}\left(\mathfrak{g}^{\otimes k}\right)$$
by setting
$$\mathsf{ad}^{(k)}_{\mathsf{X}} (\mathsf{X}_1 \otimes \cdots \otimes \mathsf{X}_k) := \sum_{i=1}^{k} \mathsf{X}_1 \otimes \cdots \otimes \mathsf{ad}_{\mathsf{X}} \mathsf{X}_i \otimes \cdots \otimes \mathsf{X}_k,$$
with $\mathsf{X} \in \mathfrak{g}$. The Jacobi identity in $\mathfrak{g}$ implies that $\mathsf{ad}^{(k)}$ is a representation of $\mathfrak{g}$ on $\mathscr{T}(\mathfrak{g})$.

- Fix $k = 2$ (we will be interested in such a case in Chapter 4).

  (a) Consider $\mathsf{X}, \mathsf{Y}, \mathsf{Z} \in \mathfrak{g}$. Then
$$\begin{aligned} \mathsf{ad}^{(2)}_{\mathsf{X}} (\mathsf{Y} \otimes \mathsf{Z}) &:= (\mathsf{ad}_{\mathsf{X}} \mathsf{Y}) \otimes \mathsf{Z} + \mathsf{Y} \otimes (\mathsf{ad}_{\mathsf{X}} \mathsf{Z}) \\ &= [\mathsf{X}, \mathsf{Y}] \otimes \mathsf{Z} + \mathsf{Y} \otimes [\mathsf{X}, \mathsf{Z}]. \end{aligned}$$
  If 1 denotes the identity map from $\mathfrak{g}$ to $\mathfrak{g}$ we write
$$\mathsf{ad}^{(2)}_{\mathsf{X}} = \mathsf{ad}_{\mathsf{X}} \otimes 1 + 1 \otimes \mathsf{ad}_{\mathsf{X}}.$$
  The map $\mathsf{ad}^{(2)} : \mathfrak{g} \to \mathsf{End}(\mathfrak{g} \otimes \mathfrak{g})$ provides a representation of $\mathfrak{g}$ on $\mathfrak{g} \otimes \mathfrak{g}$.

  (b) Consider the following space of ℓ-cochains of $\mathfrak{g}$:
$$\mathcal{C}^{\ell}(\mathfrak{g}, \mathfrak{g} \otimes \mathfrak{g}) := \mathsf{Hom}\left(\mathfrak{g}^{\wedge \ell}, \mathfrak{g} \otimes \mathfrak{g}\right).$$
  Then $\mathcal{C}^{\ell}(\mathfrak{g}, \mathfrak{g} \otimes \mathfrak{g})$ is equipped with a coboundary operator
$$\delta : \mathcal{C}^{\ell}(\mathfrak{g}, \mathfrak{g} \otimes \mathfrak{g}) \to \mathcal{C}^{\ell+1}(\mathfrak{g}, \mathfrak{g} \otimes \mathfrak{g}).$$
  If $r \in \mathfrak{g} \otimes \mathfrak{g} \simeq \mathcal{C}^{0}(\mathfrak{g}, \mathfrak{g} \otimes \mathfrak{g})$ we can define a linear map $\gamma := \delta r : \mathfrak{g} \to \mathfrak{g} \otimes \mathfrak{g}$ given by (see (1.35) and Example 1.20)
$$\gamma(\mathsf{X}) := \delta r(\mathsf{X}) = \mathsf{ad}^{(2)}_{\mathsf{X}} r \tag{1.44}$$
  for all $\mathsf{X} \in \mathfrak{g}$.

1.3.5 Functions invariant under adjoint actions

► A function $H \in \mathscr{F}(\mathfrak{g})$ is called Ad-*invariant* if

$$H(\mathsf{Ad}_h \mathsf{X}) = H(\mathsf{X}) \qquad \forall h \in \mathbf{G}, \mathsf{X} \in \mathfrak{g}.$$

Similarly, a function $H \in \mathscr{F}(\mathfrak{g}^*)$ is called Ad*-*invariant* if

$$H(\mathsf{Ad}_h^* \mathsf{X}^*) = H(\mathsf{X}^*) \qquad \forall h \in \mathbf{G}, \mathsf{X}^* \in \mathfrak{g}^*. \tag{1.45}$$

The algebra of Ad-invariant functions on $\mathfrak{g}$ is denoted by $\mathscr{F}_{\mathbf{G}}(\mathfrak{g})$, while the algebra of Ad*-invariant functions on $\mathfrak{g}^*$ is denoted by $\mathscr{F}_{\mathbf{G}}(\mathfrak{g}^*)$. It is useful to observe that if $H \in \mathscr{F}(\mathfrak{g}^*)$ then the differential of H at $\mathsf{X}^* \in \mathfrak{g}^*$ is a linear map $\mathsf{d}H(\mathsf{X}^*) : T_{\mathsf{X}^*}\mathfrak{g}^* \to \mathbb{R}$, which can be identified with an element of $\mathfrak{g}$.

► The following claim describes two fundamental properties of Ad*-invariant functions. Ad-invariant functions have similar properties.

Theorem 1.1

Let $H \in \mathscr{F}_{\mathbf{G}}(\mathfrak{g}^*)$.

1. *For any $\mathsf{X}^* \in \mathfrak{g}^*$ and for any $\mathsf{X} \in \mathfrak{g}$ we have*

$$\langle \mathsf{X}^*, [\mathsf{d}H(\mathsf{X}^*), \mathsf{X}] \rangle = 0, \tag{1.46}$$

i.e., for any $\mathsf{X}^* \in \mathfrak{g}^*$ we have

$$\mathsf{ad}^*_{\mathsf{d}H(\mathsf{X}^*)} \mathsf{X}^* = 0. \tag{1.47}$$

2. *For any $\mathsf{X}^* \in \mathfrak{g}^*$ and for any $h \in \mathbf{G}$ the following diagram is commutative:*

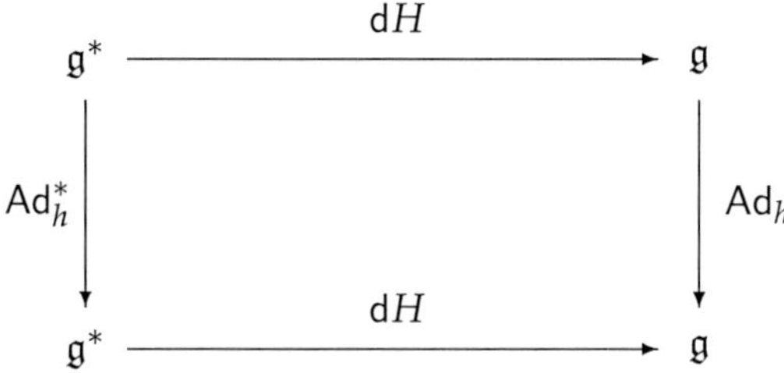

In formulas,

$$\mathsf{d}H(\mathsf{Ad}_h^* \mathsf{X}^*) = \mathsf{Ad}_h(\mathsf{d}H(\mathsf{X}^*)).$$

Proof. We prove both claims.

1. Let $\mathsf{X} \in \mathfrak{g}$ and $\mathsf{X}^* \in \mathfrak{g}^*$. If $H \in \mathscr{F}_{\mathbf{G}}(\mathfrak{g}^*)$ then $H(\mathsf{Ad}_h^* \mathsf{X}^*) = H(\mathsf{X}^*)$ for all $\mathsf{X}^* \in \mathfrak{g}^*$ and $h \in \mathbf{G}$. Taking any $\mathsf{X} \in \mathfrak{g}$ we have (use (1.41), (1.4), (1.42) and (1.45)):

$$\begin{aligned}\left\langle \mathsf{ad}^*_{\mathsf{d}H(\mathsf{X}^*)} \mathsf{X}^*, \mathsf{X} \right\rangle &= -\langle \mathsf{ad}^*_{\mathsf{X}} \mathsf{X}^*, \mathsf{d}H(\mathsf{X}^*) \rangle \\ &= -\mathsf{ad}^*_{\mathsf{X}} \mathsf{X}^*[H(\mathsf{X}^*)] \\ &= -\left.\frac{\mathsf{d}}{\mathsf{d}t}\right|_{t=0} H\left(\mathsf{Ad}^*_{\exp(t\mathsf{X})} \mathsf{X}^*\right) \\ &= -\left.\frac{\mathsf{d}}{\mathsf{d}t}\right|_{t=0} H(\mathsf{X}^*) = 0.\end{aligned}$$

Equivalence between (1.46) and (1.47) is a consequence of (1.40).

2. In order to prove that the above diagram is commutative, differentiate for a fixed $h \in \mathbf{G}$ the identity $H = H \circ \mathsf{Ad}_h^*$ at $\mathsf{X}^* \in \mathfrak{g}^*$. It gives

$$\mathsf{d}H(\mathsf{X}^*) = \mathsf{d}H(\mathsf{Ad}_h^* \mathsf{X}^*) \circ (\mathsf{d}\mathsf{Ad}_h^*)(\mathsf{X}^*) = \mathsf{d}H(\mathsf{Ad}_h^* \mathsf{X}^*) \circ \mathsf{Ad}_h^*,$$

because, for fixed $h \in \mathbf{G}$, the map Ad_h^* is a linear map. Thus, for $\mathsf{Y}^* \in \mathfrak{g}^*$, we have:

$$\begin{aligned}\langle \mathsf{d}H(\mathsf{X}^*), \mathsf{Y}^* \rangle &= \langle \mathsf{d}H(\mathsf{Ad}_h^* \mathsf{X}^*), \mathsf{Ad}_h^* \mathsf{Y}^* \rangle \\ &= \langle \mathsf{Ad}_{h^{-1}} (\mathsf{d}H(\mathsf{Ad}_h^* \mathsf{X}^*)), \mathsf{Y}^* \rangle,\end{aligned}$$

where we used (1.39). Then

$$\mathsf{d}H(\mathsf{X}^*) = \mathsf{Ad}_{h^{-1}} (\mathsf{d}H(\mathsf{Ad}_h^* \mathsf{X}^*)).$$

It follows that $\mathsf{d}H(\mathsf{Ad}_h^* \mathsf{X}^*) = \mathsf{Ad}_h(\mathsf{d}H(\mathsf{X}^*))$, which proves that the diagram is commutative.

The Theorem is proved. ∎

1.3.6 The Killing form

► Lie algebras often are equipped with a non-degenerate symmetric bilinear form

$$\langle \cdot | \cdot \rangle : \mathfrak{g} \times \mathfrak{g} \to \mathbb{R}.$$

- Such a form allows us to identify $\mathfrak{g}$ with $\mathfrak{g}^*$ simply by defining an isomorphism $\sigma : \mathfrak{g} \to \mathfrak{g}^*$ that assigns to $\mathsf{X} \in \mathfrak{g}$ the linear form $\sigma(\mathsf{X}) \in \mathfrak{g}^*$ according to the rule

$$\langle \sigma(\mathsf{X}), \mathsf{Y} \rangle = \langle \mathsf{X} | \mathsf{Y} \rangle \qquad \forall \mathsf{X}, \mathsf{Y} \in \mathfrak{g}. \tag{1.48}$$

- The inverse of σ is the linear map $\sigma^{-1} : \mathfrak{g}^* \to \mathfrak{g} : \mathsf{X}^* \mapsto \mathsf{X}$, where X is the unique element of $\mathfrak{g}$ that satisfies

$$\left\langle \sigma^{-1}(\mathsf{X}^*) \,\middle|\, \mathsf{Y} \right\rangle = \langle \mathsf{X} \,|\, \mathsf{Y} \rangle = \langle \mathsf{X}^*, \mathsf{Y} \rangle \qquad \forall \mathsf{Y} \in \mathfrak{g}.$$

This allows us to write $\mathfrak{g} \simeq \mathfrak{g}^*$. Abusing the notation we can also write $\langle \cdot \,|\, \cdot \rangle \equiv \langle \cdot, \cdot \rangle$.

- A symmetric bilinear form $\langle \cdot \,|\, \cdot \rangle$ on $\mathfrak{g}$ is called Ad-*invariant* if for any $h \in \mathbf{G}$ and for any $\mathsf{X}, \mathsf{Y} \in \mathfrak{g}$ one has

$$\langle \mathsf{Ad}_h \mathsf{X} \,|\, \mathsf{Ad}_h \mathsf{Y} \rangle = \langle \mathsf{X} \,|\, \mathsf{Y} \rangle .$$

- A Lie algebra equipped with a non-degenerate symmetric bilinear form which is Ad-invariant is called *quadratic Lie algebra*.

- Ad-invariance of $\langle \cdot \,|\, \cdot \rangle$ implies the following properties:

  1. There holds:

$$\langle \mathsf{ad}_\mathsf{Y} \mathsf{X} \,|\, \mathsf{Z} \rangle = - \langle \mathsf{X} \,|\, \mathsf{ad}_\mathsf{Y} \mathsf{Z} \rangle \qquad \forall \mathsf{X}, \mathsf{Y}, \mathsf{Z} \in \mathfrak{g}, \tag{1.49}$$

  which means that ad_Y is skew-symmetric w.r.t. $\langle \cdot \,|\, \cdot \rangle$.

  2. There holds:

$$\mathsf{Ad}_h^* \sigma(\mathsf{X}) = \sigma(\mathsf{Ad}_h \mathsf{X}), \qquad \forall h \in \mathbf{G}, \mathsf{X} \in \mathfrak{g}, \tag{1.50}$$

  or, equivalently, the following diagram is commutative for any $h \in \mathbf{G}$:

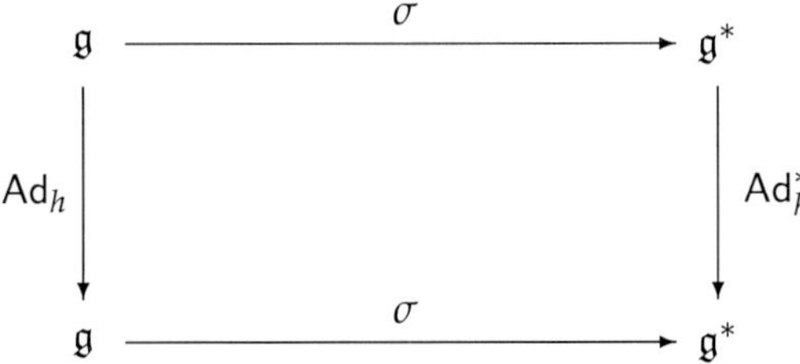

  Indeed for any $\mathsf{X}, \mathsf{Y} \in \mathfrak{g}$ it follows that

$$\begin{aligned} \langle \mathsf{Ad}_h^* \sigma(\mathsf{X}), \mathsf{Y} \rangle &= \langle \sigma(\mathsf{X}), \mathsf{Ad}_{h^{-1}} \mathsf{Y} \rangle = \langle \mathsf{X} \,|\, \mathsf{Ad}_{h^{-1}} \mathsf{Y} \rangle \\ &= \langle \mathsf{Ad}_h \mathsf{X} \,|\, \mathsf{Y} \rangle = \langle \sigma(\mathsf{Ad}_h \mathsf{X}), \mathsf{Y} \rangle . \end{aligned}$$

  In other words, upon identifying a Lie algebra with its dual the adjoint and coadjoint actions are identified.

Example 1.22 (*The Lie algebra* $\mathfrak{so}(3)$)

The Lie algebra $\mathfrak{so}(3)$ of skew-symmetric 3×3 matrices is isomorphic to $\mathbb{R}^3$. Indeed there exists an isomorphism φ such that $(\mathfrak{so}(3), [\cdot,\cdot]) \simeq (\mathbb{R}^3, \times)$. It is explicitly given by

$$x := (x_1, x_2, x_3)^\top \overset{\varphi}{\longleftrightarrow} \mathsf{X} := \begin{pmatrix} 0 & -x_3 & x_2 \\ x_3 & 0 & -x_1 \\ -x_2 & x_1 & 0 \end{pmatrix}.$$

- If $\mathsf{X} \in \mathfrak{so}(3)$ the adjoint action of $\mathbf{SO}(3)$ is given by

$$\mathsf{Ad}_h \mathsf{X} = h \mathsf{X} h^{-1}, \qquad h \in \mathbf{SO}(3).$$

  One can check that for all matrices $h \in \mathbf{SO}(3)$ one has

$$\varphi(h\,x) = h\,\varphi(x)\,h^{-1}, \qquad x \in \mathbb{R}^3,$$

  so that the adjoint action of $\mathbf{SO}(3)$ is identified with the action of $\mathbf{SO}(3)$ on $\mathbb{R}^3$ given by $(h, x) \mapsto h\,x$.
- The Euclidean scalar product on $\mathbb{R}^3$ defines a non-degenerate symmetric bilinear form on $\mathbb{R}^3 \simeq \mathfrak{so}(3)$. By using the condition $h^\top h = 1_3$ one finds that this form is Ad-invariant. Thus, we may use the scalar product to identify the coadjoint and adjoint actions of $\mathbf{SO}(3)$.
- If $x \in \mathbb{R}^3$ we have

$$\mathsf{Ad}^*_h\, x = h\, x,$$

  and therefore the coadjoint orbit through x is:

$$\mathcal{O}_x = \{\mathsf{Ad}^*_h\, x \,:\, h \in \mathbf{SO}(3)\} = \{h\,x \,:\, h \in \mathbf{SO}(3)\}.$$

  If $x \neq 0$ the set $\mathcal{O}_x$ is a sphere of radius $\|x\|$, while if $x = 0$ the set $\mathcal{O}_x$ is a single point.

► The most important example of an Ad-invariant symmetric bilinear form (not necessarily non-degenerate) is the *Killing form* of $\mathfrak{g}$, which is defined by

$$\langle \mathsf{X} \,|\, \mathsf{Y} \rangle := \mathsf{Trace}\,(\mathsf{ad}_\mathsf{X}\, \mathsf{ad}_\mathsf{Y}), \qquad \mathsf{X}, \mathsf{Y} \in \mathfrak{g}. \tag{1.51}$$

► We give the following statement.

Theorem 1.2

1. *The Killing form is symmetric, i.e.,* $\langle \mathsf{X} \,|\, \mathsf{Y} \rangle = \langle \mathsf{Y} \,|\, \mathsf{X} \rangle$ *for all* $\mathsf{X}, \mathsf{Y} \in \mathfrak{g}$.
2. *The Killing form is* Ad*-invariant, i.e.,*

$$\langle \mathsf{X} \,|\, \mathsf{Y} \rangle = \langle \mathsf{Ad}_h \mathsf{X} \,|\, \mathsf{Ad}_h \mathsf{Y} \rangle, \qquad \forall h \in \mathbf{G},\ \mathsf{X}, \mathsf{Y} \in \mathfrak{g}.$$

3. *The Killing form satisfies (1.49), i.e.,*

$$\langle [\mathsf{Y}, \mathsf{X}] \,|\, \mathsf{Z} \rangle = -\langle \mathsf{X} \,|\, [\mathsf{Y}, \mathsf{Z}] \rangle \qquad \forall \mathsf{X}, \mathsf{Y}, \mathsf{Z} \in \mathfrak{g}.$$

Proof. We prove all claims.

1. It follows from the symmetry of the trace.
2. Let us assume that **G** is a matrix Lie group. We have:

$$\begin{aligned}\mathrm{ad}_{\mathrm{Ad}_h \mathsf{X}} \mathsf{Y} &= [\,\mathrm{Ad}_h \mathsf{X}, \mathsf{Y}\,] \\ &= \left[\mathrm{Ad}_h \mathsf{X}, \mathrm{Ad}_h \mathrm{Ad}_h^{-1} \mathsf{Y}\right] \\ &= \mathrm{Ad}_h \left[\mathsf{X}, \mathrm{Ad}_h^{-1} \mathsf{Y}\right] \\ &= \mathrm{Ad}_h \,\mathrm{ad}_{\mathsf{X}}\, \mathrm{Ad}_h^{-1} \mathsf{Y},\end{aligned}$$

so that

$$\mathrm{ad}_{\mathrm{Ad}_h \mathsf{X}} = \mathrm{Ad}_h \,\mathrm{ad}_{\mathsf{X}}\, \mathrm{Ad}_h^{-1}.$$

Now we have:

$$\begin{aligned}\langle\, \mathrm{Ad}_h \mathsf{X} \,|\, \mathrm{Ad}_h \mathsf{Y}\,\rangle &= \mathrm{Trace}(\mathrm{ad}_{\mathrm{Ad}_h \mathsf{X}}\, \mathrm{ad}_{\mathrm{Ad}_h \mathsf{Y}}) \\ &= \mathrm{Trace}\left(\mathrm{Ad}_h \,\mathrm{ad}_{\mathsf{X}}\, \mathrm{Ad}_h^{-1}\, \mathrm{Ad}_h \,\mathrm{ad}_{\mathsf{Y}}\, \mathrm{Ad}_h^{-1}\right) \\ &= \mathrm{Trace}\left(\mathrm{Ad}_h \,\mathrm{ad}_{\mathsf{X}}\, \mathrm{ad}_{\mathsf{Y}}\, \mathrm{Ad}_h^{-1}\right) \\ &= \mathrm{Trace}\,(\mathrm{ad}_{\mathsf{X}}\, \mathrm{ad}_{\mathsf{Y}}) \\ &= \langle \mathsf{X} \,|\, \mathsf{Y}\,\rangle,\end{aligned}$$

where we used the conjugation invariance of the trace.

3. We compute, using (1.38),

$$\begin{aligned}\langle\, [\,\mathsf{Y},\mathsf{X}\,] \,|\, \mathsf{Z}\,\rangle &= \mathrm{Trace}\left(\mathrm{ad}_{[\,\mathsf{Y},\mathsf{X}\,]}\, \mathrm{ad}_{\mathsf{Z}}\right) \\ &= \mathrm{Trace}\,(\mathrm{ad}_{\mathsf{Y}}\, \mathrm{ad}_{\mathsf{X}}\, \mathrm{ad}_{\mathsf{Z}} - \mathrm{ad}_{\mathsf{X}}\, \mathrm{ad}_{\mathsf{Y}}\, \mathrm{ad}_{\mathsf{Z}}) \\ &= \mathrm{Trace}\,(\mathrm{ad}_{\mathsf{X}}\, \mathrm{ad}_{\mathsf{Z}}\, \mathrm{ad}_{\mathsf{Y}} - \mathrm{ad}_{\mathsf{X}}\, \mathrm{ad}_{\mathsf{Y}}\, \mathrm{ad}_{\mathsf{Z}}) \\ &= \mathrm{Trace}\left(\mathrm{ad}_{\mathsf{X}}\, \mathrm{ad}_{[\,\mathsf{Z},\mathsf{Y}\,]}\right) \\ &= \langle \mathsf{X} \,|\, [\,\mathsf{Z},\mathsf{Y}\,]\rangle \\ &= -\langle \mathsf{X} \,|\, [\,\mathsf{Y},\mathsf{Z}\,]\,\rangle,\end{aligned}$$

where in passing to the third line we have used the invariance of the trace under cyclic permutations of order.

The Theorem is proved. ■

► Without giving details about the problem of Lie algebras classification we just provide the following definitions and notions.

- A non-empty subset $\mathfrak{t}$ of a Lie algebra $\mathfrak{g}$ is called an *ideal* if $[\,\mathfrak{g}, \mathfrak{t}\,] \subseteq \mathfrak{t}$.

- If $\mathfrak{g}$ contains no other ideals than the empty set and itself then $\mathfrak{g}$ is a *simple Lie algebra*. A Lie algebra that is isomorphic to the direct sum of simple Lie algebras is called a *semi-simple Lie algebra*.
- Semi-simple (and simple) Lie algebras are characterized by the fact that their Killing form is non-degenerate ("Cartan criterion").
- For any simple Lie algebra $\mathfrak{g}$ the Killing form is, up to a constant, the unique Ad-invariant symmetric bilinear form on $\mathfrak{g}$ that is non-degenerate.

Example 1.23 (*The Lie algebra* $\mathfrak{e}(3)$)

Consider the six-dimensional real Lie algebra $\mathfrak{e}(3) \simeq \mathfrak{so}(3) \oplus_s \mathbb{R}^3$ ($\oplus_s$ denotes the semi-direct sum of vector spaces) corresponding to the Euclidean group $\mathbf{E}(3)$ defined by rotations and translations.

- Let $\mathsf{E}_1, \mathsf{E}_2, \mathsf{E}_3$ be the infinitesimal generators of rotations and by $\mathsf{D}_1, \mathsf{D}_2, \mathsf{D}_3$ be the infinitesimal generators of translations. Then $\{\mathsf{E}_1, \mathsf{E}_2, \mathsf{E}_3, \mathsf{D}_1, \mathsf{D}_2, \mathsf{D}_3\}$ is a basis of $\mathfrak{e}(3)$ and the Lie brackets are:
$$[\mathsf{E}_i, \mathsf{E}_j] = \varepsilon_{ijk}\mathsf{E}_k, \qquad [\mathsf{E}_i, \mathsf{D}_j] = \varepsilon_{ijk}\mathsf{D}_k, \qquad [\mathsf{D}_i, \mathsf{D}_j] = 0.$$
- The Abelian Lie algebra $\mathbb{R}^3$, spanned by $\mathsf{D}_1, \mathsf{D}_2, \mathsf{D}_3$ is an ideal of $\mathfrak{e}(3)$. But $\mathfrak{e}(3)$ is not semi-simple due to its semi-direct sum structure.

Example 1.24 (*The Killing form of* $\mathfrak{sl}(2, \mathbb{R})$)

Consider $\mathfrak{sl}(2, \mathbb{R}) := \{\mathsf{X} \in \mathfrak{gl}(2, \mathbb{R}) \,:\, \mathsf{Trace}\,\mathsf{X} = 0\}$ with basis
$$\mathsf{E}_1 := \begin{pmatrix} 0 & 1 \\ 0 & 0 \end{pmatrix}, \qquad \mathsf{E}_2 := \begin{pmatrix} 1 & 0 \\ 0 & -1 \end{pmatrix}, \qquad \mathsf{E}_3 := \begin{pmatrix} 0 & 0 \\ 1 & 0 \end{pmatrix},$$
and Lie brackets
$$[\mathsf{E}_1, \mathsf{E}_2] = -2\,\mathsf{E}_1, \qquad [\mathsf{E}_1, \mathsf{E}_3] = \mathsf{E}_2, \qquad [\mathsf{E}_3, \mathsf{E}_2] = 2\,\mathsf{E}_3.$$

- By direct computation we have:
$$\mathsf{ad}_{\mathsf{E}_1} \mathsf{E}_1 = [\mathsf{E}_1, \mathsf{E}_1] = 0, \qquad \mathsf{ad}_{\mathsf{E}_1} \mathsf{E}_2 = [\mathsf{E}_1, \mathsf{E}_2] = -2\,\mathsf{E}_1, \qquad \mathsf{ad}_{\mathsf{E}_1} \mathsf{E}_3 = [\mathsf{E}_1, \mathsf{E}_3] = \mathsf{E}_2,$$
$$\mathsf{ad}_{\mathsf{E}_2} \mathsf{E}_1 = [\mathsf{E}_2, \mathsf{E}_1] = 2\,\mathsf{E}_1, \qquad \mathsf{ad}_{\mathsf{E}_2} \mathsf{E}_2 = [\mathsf{E}_2, \mathsf{E}_2] = 0, \qquad \mathsf{ad}_{\mathsf{E}_2} \mathsf{E}_3 = [\mathsf{E}_2, \mathsf{E}_3] = -2\,\mathsf{E}_3,$$
and
$$\mathsf{ad}_{\mathsf{E}_3} \mathsf{E}_1 = [\mathsf{E}_3, \mathsf{E}_1] = -\mathsf{E}_2, \qquad \mathsf{ad}_{\mathsf{E}_3} \mathsf{E}_2 = [\mathsf{E}_3, \mathsf{E}_2] = 2\,\mathsf{E}_3, \qquad \mathsf{ad}_{\mathsf{E}_3} \mathsf{E}_3 = [\mathsf{E}_3, \mathsf{E}_3] = 0.$$
Therefore
$$\mathsf{ad}_{\mathsf{E}_1} = \begin{pmatrix} 0 & -2 & 0 \\ 0 & 0 & 1 \\ 0 & 0 & 0 \end{pmatrix}, \quad \mathsf{ad}_{\mathsf{E}_2} = \begin{pmatrix} 2 & 0 & 0 \\ 0 & 0 & 0 \\ 0 & 0 & -2 \end{pmatrix}, \quad \mathsf{ad}_{\mathsf{E}_3} = \begin{pmatrix} 0 & 0 & 0 \\ -1 & 0 & 0 \\ 0 & 2 & 0 \end{pmatrix}.$$
Note that
$$[\mathsf{ad}_{\mathsf{E}_1}, \mathsf{ad}_{\mathsf{E}_2}] = -2\,\mathsf{ad}_{\mathsf{E}_1}, \quad [\mathsf{ad}_{\mathsf{E}_1}, \mathsf{ad}_{\mathsf{E}_3}] = \mathsf{ad}_{\mathsf{E}_2}, \quad [\mathsf{ad}_{\mathsf{E}_3}, \mathsf{ad}_{\mathsf{E}_2}] = 2\,\mathsf{ad}_{\mathsf{E}_3}.$$
- The matrix associated with the Killing form is
$$A := \begin{pmatrix} \langle \mathsf{E}_1 \,|\, \mathsf{E}_1 \rangle & \langle \mathsf{E}_1 \,|\, \mathsf{E}_2 \rangle & \langle \mathsf{E}_1 \,|\, \mathsf{E}_3 \rangle \\ \langle \mathsf{E}_2 \,|\, \mathsf{E}_1 \rangle & \langle \mathsf{E}_2 \,|\, \mathsf{E}_2 \rangle & \langle \mathsf{E}_2 \,|\, \mathsf{E}_3 \rangle \\ \langle \mathsf{E}_3 \,|\, \mathsf{E}_1 \rangle & \langle \mathsf{E}_3 \,|\, \mathsf{E}_2 \rangle & \langle \mathsf{E}_3 \,|\, \mathsf{E}_3 \rangle \end{pmatrix} = \begin{pmatrix} 0 & 0 & 4 \\ 0 & 8 & 0 \\ 4 & 0 & 0 \end{pmatrix},$$
which has rank 3. Therefore the Killing form is non-degenerate and $\mathfrak{sl}(2, \mathbb{R})$ is simple.

- We can identify $\mathfrak{sl}(2,\mathbb{R})$ with its dual $\mathfrak{sl}(2,\mathbb{R})^*$ by means of the Killing form. Indeed if we denote by $\mathsf{E}_1^*, \mathsf{E}_2^*, \mathsf{E}_3^*$ the basis $\mathfrak{sl}(2,\mathbb{R})^*$ with $\left\langle \mathsf{E}_i, \mathsf{E}_j^* \right\rangle = \delta_{ij}$, we have

$$\mathsf{E}_i^* = \sum_{j=1}^{3} \left(A^{-1}\right)_{ij} \mathsf{E}_j,$$

which gives

$$\mathsf{E}_1^* = \frac{1}{4}\,\mathsf{E}_3, \qquad \mathsf{E}_2^* = \frac{1}{8}\,\mathsf{E}_2, \qquad \mathsf{E}_3^* = \frac{1}{4}\,\mathsf{E}_1.$$

It turns out that

$$[\,\mathsf{E}_1^*, \mathsf{E}_2^*\,] = \frac{1}{4}\,\mathsf{E}_1^*, \qquad [\,\mathsf{E}_1^*, \mathsf{E}_3^*\,] = -\frac{1}{2}\,\mathsf{E}_2^*, \qquad [\,\mathsf{E}_3^*, \mathsf{E}_2^*\,] = -\frac{1}{4}\,\mathsf{E}_3^*,$$

which is obviously isomorphic to $\mathfrak{sl}(2,\mathbb{R})$.

Example 1.25 (*The Killing form of* $\mathfrak{gl}(n,\mathbb{R})$, $\mathfrak{sl}(n,\mathbb{R})$ *and* $\mathfrak{so}(n)$)

- Let us construct the Killing form for the Lie algebra $\mathfrak{gl}(n,\mathbb{R})$. It is clear that the $n \times n$ matrices E_{ij}, $i,j = 1,\dots,n$, with matrix elements $(\mathsf{E}_{ij})_{k\ell} := \delta_{ik}\,\delta_{j\ell}$ form a basis for $\mathfrak{gl}(n,\mathbb{R})$. One shows that $\mathsf{E}_{ij}\,\mathsf{E}_{k\ell} = \delta_{jk}\mathsf{E}_{i\ell}$ from which the following Lie brackets follow:

$$[\,\mathsf{E}_{ij}, \mathsf{E}_{k\ell}\,] = \delta_{jk}\,\mathsf{E}_{i\ell} - \delta_{\ell i}\,\mathsf{E}_{kj}. \tag{1.52}$$

- For any element $\mathsf{X} \in \mathfrak{gl}(n,\mathbb{R})$ we have the representation

$$\mathsf{X} = \sum_{i,j=1}^{n} \mathsf{X}_{ij}\,\mathsf{E}_{ij}.$$

for some smooth coordinates X_{ij} which form a functional basis of $\mathscr{F}(\mathfrak{gl}(n,\mathbb{R}))$.

- Taking into account (1.52) one obtains

$$\mathrm{ad}_{\mathsf{X}}\,\mathsf{E}_{ij} = \sum_{k,\ell=1}^{n} (\mathsf{X}_{ki}\,\delta_{\ell j} - \mathsf{X}_{j\ell}\,\delta_{ki})\,\mathsf{E}_{k\ell},$$

which gives

$$\mathrm{ad}_{\mathsf{X}}\,\mathrm{ad}_{\mathsf{Y}}\,\mathsf{E}_{ij} = \sum_{k,r,s=1}^{n} (\mathsf{X}_{rk}\,\mathsf{Y}_{ki}\,\delta_{sj} + \mathsf{X}_{ks}\,\mathsf{Y}_{jk}\,\delta_{ri} - \mathsf{X}_{ri}\,\mathsf{Y}_{jk}\,\delta_{sk} - \mathsf{X}_{js}\,\mathsf{Y}_{ki}\,\delta_{rk})\,\mathsf{E}_{rs},$$

for

$$\mathsf{Y} = \sum_{i,j=1}^{n} \mathsf{Y}_{ij}\,\mathsf{E}_{ij}.$$

- Then a direct computation gives

$$\langle\,\mathsf{X}\,|\,\mathsf{Y}\,\rangle = 2\,n\,\mathrm{Trace}\,(\mathsf{X}\,\mathsf{Y}) - 2\,\mathrm{Trace}\,\mathsf{X}\,\mathrm{Trace}\,\mathsf{Y}.$$

- For the special Lie algebra $\mathfrak{sl}(n,\mathbb{R}) := \{\mathsf{X} \in \mathfrak{gl}(n,\mathbb{R}) \;:\; \mathrm{Trace}\,\mathsf{X} = 0\}$ the Killing form is

$$\langle\,\mathsf{X}\,|\,\mathsf{Y}\,\rangle = 2\,n\,\mathrm{Trace}\,(\mathsf{X}\,\mathsf{Y}),$$

- For the orthogonal Lie algebra $\mathfrak{so}(n) := \{\mathsf{X} \in \mathfrak{gl}(n,\mathbb{R}) \;:\; \mathsf{X} + \mathsf{X}^{\top} = 0_n\}$ the Killing form is

$$\langle\,\mathsf{X}\,|\,\mathsf{Y}\,\rangle = (n-2)\,\mathrm{Trace}\,(\mathsf{X}\,\mathsf{Y}).$$

1.3.7 Loop algebras

► A very important role in the theory of integrable systems is played by *loop algebras*, also called *affine Lie algebras*. They are infinite-dimensional Lie algebras of Laurent polynomials over some finite-dimensional Lie algebra.

- Let $\mathfrak{g}$ be a finite-dimensional Lie algebra. Then the corresponding loop algebra is
$$\mathfrak{g}\left[\lambda,\lambda^{-1}\right] := \left\{ \mathsf{X}(\lambda) = \sum_i \lambda^i \mathsf{X}_i \,:\, \mathsf{X}_i \in \mathfrak{g},\, \lambda \in \mathbb{C} \right\},$$
namely the Lie algebra of Laurent polynomials in λ with coefficients in $\mathfrak{g}$.

- The Lie bracket on $\mathfrak{g}\left[\lambda,\lambda^{-1}\right]$ is defined by setting
$$\left[\lambda^i \mathsf{X}, \lambda^j \mathsf{Y}\right] = \lambda^{i+j}\left[\mathsf{X},\mathsf{Y}\right], \qquad \mathsf{X},\mathsf{Y} \in \mathfrak{g}.$$
This means that $\mathfrak{g}\left[\lambda,\lambda^{-1}\right]$ has a natural grading by powers of λ. One formally writes
$$\mathfrak{g}\left[\lambda,\lambda^{-1}\right] = \bigoplus_i \lambda^i \mathfrak{g}.$$

- If $\mathfrak{g}$ has non-degenerate Killing form $\langle \cdot \,|\, \cdot \rangle : \mathfrak{g} \to \mathfrak{g}$, then there exists an infinite family of non-degenerate Ad-invariant bilinear forms $\langle \cdot \,|\, \cdot \rangle_i$ on $\mathfrak{g}\left[\lambda,\lambda^{-1}\right]$, enumerated by the integer i. One writes $\langle \mathsf{X}(\lambda) \,|\, \mathsf{Y}(\lambda) \rangle_i$, for $\mathsf{X}(\lambda), \mathsf{Y}(\lambda) \in \mathfrak{g}\left[\lambda,\lambda^{-1}\right]$, to denote the coefficient by λ^i of the Laurent polynomial obtained by the point-wise application of the Killing form on $\mathfrak{g}$.

2

Poisson Structures

2.1 Poisson manifolds

▶ We now have all necessary ingredients to define those smooth manifolds where Hamiltonian dynamics can be locally defined in a natural way. Our strategy will be to define the smooth manifolds (which play the role of phase spaces) on which Hamiltonian vector fields can be locally constructed and then to derive consequences in a systematic way.

▶ In what follows $\mathcal{M}$ is a real n-dimensional smooth manifold. We now give two fundamental definitions.

Definition 2.1

A **Poisson bracket** *(or* **Poisson structure***) on $\mathcal{M}$ is a Lie algebra structure $\{\cdot,\cdot\}$ on $\mathscr{F}(\mathcal{M})$, which is a biderivation on $\mathscr{F}(\mathcal{M})$. Equivalently, for any $H \in \mathscr{F}(\mathcal{M})$ the linear map*

$$\mathscr{F}(\mathcal{M}) \to \mathscr{F}(\mathcal{M}) \,:\, F \mapsto \{F,H\},$$

is a derivation on $\mathscr{F}(\mathcal{M})$, i.e., it defines a smooth vector field on $\mathcal{M}$. The pair $(\mathcal{M},\{\cdot,\cdot\})$ is called **Poisson manifold**.

Definition 2.2

Let $(\mathcal{M},\{\cdot,\cdot\})$ be a Poisson manifold.

1. *For any $H \in \mathscr{F}(\mathcal{M})$ the vector field*

$$\mathsf{v}_H := \{\cdot,H\} \tag{2.1}$$

is called **Hamiltonian vector field** *associated with the* **Hamiltonian** *H. We write*

$$\mathsf{Ham}(\mathcal{M}) := \{\{\cdot,H\} \,:\, H \in \mathscr{F}(\mathcal{M})\}$$

for the vector space of Hamiltonian vector fields. The local flow $\Phi_t : \mathcal{I}_\varepsilon \times \mathcal{M} \to \mathcal{M}$, $\mathcal{I}_\varepsilon := \{t \in \mathbb{R} \,:\, |t| < \varepsilon\}$, of $\mathsf{v}_H \in \mathsf{Ham}(\mathcal{M})$ is called **Hamiltonian flow**.

2. *A function $H \in \mathscr{F}(\mathcal{M})$ is a* **Casimir** *(or* **Casimir function***) on $(\mathcal{M},\{\cdot,\cdot\})$ if its associated vector field is zero, $\mathsf{v}_H = 0$, i.e., $\{F,H\} = 0$ for all $F \in \mathscr{F}(\mathcal{M})$. We write*

$$\mathsf{Cas}(\mathcal{M}) := \{H \in \mathscr{F}(\mathcal{M}) \,:\, \mathsf{v}_H = 0\}$$

for the algebra of Casimir functions.

► From the above definitions we can systematically derive the general features of Hamiltonian dynamics on Poisson manifolds.

- Definition 2.1 says that a Poisson bracket on $\mathcal{M}$ is a bilinear operation

$$\{\cdot,\cdot\} : \mathscr{F}(\mathcal{M}) \times \mathscr{F}(\mathcal{M}) \to \mathscr{F}(\mathcal{M})$$

  possessing the following properties:

  1. *Skew-symmetry*: $\{F,G\} = -\{G,F\}$,
  2. *Leibniz rule*: $\{FG,H\} = F\{G,H\} + G\{F,H\}$,
  3. *Jacobi identity*: $\{F,\{G,H\}\} + \circlearrowleft (F,G,H) = 0$,

  for all functions $F,G,H \in \mathscr{F}(\mathcal{M})$. The algebra of smooth functions $\mathscr{F}(\mathcal{M})$ equipped with a Poisson structure is a *Poisson algebra*.

- Let $F,G \in \mathscr{F}(\mathcal{M})$. We say that F and G are in *Poisson involution* if $\{F,G\} = 0$.

- The Leibniz property for $\{\cdot,\cdot\}$ implies on the one hand that $\mathsf{Cas}(\mathcal{M})$ is an Abelian subalgebra of $\mathscr{F}(\mathcal{M})$ (for the ordinary multiplication of functions), called *center* of the Poisson algebra, and on the other hand that $\mathsf{Ham}(\mathcal{M})$ is a $\mathsf{Cas}(\mathcal{M})$-module.

- Since the Poisson bracket is a biderivation, it vanishes whenever one of its arguments is constant. We can associate with a Poisson bracket $\{\cdot,\cdot\}$ a $\mathscr{F}(\mathcal{M})$-bilinear map

$$\begin{aligned} \mathsf{P} : \Omega(\mathcal{M}) \times \Omega(\mathcal{M}) &\to \mathscr{F}(\mathcal{M}) \\ (\mathsf{d}F, \mathsf{d}G) &\mapsto \mathsf{P}(\mathsf{d}F, \mathsf{d}G) := \{F,G\}, \end{aligned}$$

  for all $F,G \in \mathscr{F}(\mathcal{M})$. $\mathsf{P} \in \mathfrak{X}^2(\mathcal{M})$ is a 2-vector field on $\mathcal{M}$ called *Poisson 2-vector field* associated with the Poisson structure $\{\cdot,\cdot\}$. We also write $\mathsf{P}[F,G] \equiv \mathsf{P}(\mathsf{d}F,\mathsf{d}G)$.

  (a) One can contract P with a 1-form to get a vector field on $\mathcal{M}$. In this way we derive from P a map

$$\widetilde{\mathsf{P}} : \Omega(\mathcal{M}) \to \mathfrak{X}(\mathcal{M}),$$

  simply by setting

$$\widetilde{\mathsf{P}}(\mathsf{d}H) := \mathsf{P}(\cdot,\mathsf{d}H) = \{\cdot,H\} = \mathsf{v}_H \qquad \forall H \in \mathscr{F}(\mathcal{M}). \tag{2.2}$$

  Formula (2.2) is an alternative definition of a Hamiltonian vector field.

(b) Let Φ_t be the Hamiltonian flow corresponding to v_H. We say that $F \in \mathscr{F}(\mathcal{M})$ is an *integral of motion* of Φ_t (i.e., an invariant function for the group action Φ_t) if it is in Poisson involution with H, i.e.,

$$\mathsf{P}(\mathsf{d}F, \mathsf{d}H) = \{ F, H \} = \mathsf{v}_H[F] = \mathfrak{L}_{\mathsf{v}_H} F = 0.$$

In particular, H is an integral of motion of Φ_t.

(c) We see from (2.2) that a Casimir function $H \in \mathsf{Cas}(\mathcal{M})$ is such that $\widetilde{\mathsf{P}}$ annihilates its differential $\mathsf{d}H$, i.e., $\widetilde{\mathsf{P}}(\mathsf{d}H) = 0$. Equivalently $H \in \mathsf{Cas}(\mathcal{M})$ if and only if $\mathsf{d}H \in \mathsf{Ker}\,\widetilde{\mathsf{P}}$. Note that Casimir functions are constant along the integral curves of any Hamiltonian vector field. Therefore one says that Casimirs are trivial integrals of motion.

(d) A necessary and sufficient condition for a 2-vector field $\mathsf{P} \in \mathfrak{X}^2(\mathcal{M})$ to define a Poisson structure is that

$$[\{ \cdot, \cdot \}, \{ \cdot, \cdot \}]_{\mathrm{SN}} = 0, \tag{2.3}$$

where $[\,\cdot, \cdot\,]_{\mathrm{SN}}$, called *Schouten-Nijenhuis bracket* between two Poisson brackets $\{ \cdot, \cdot \}_1$ and $\{ \cdot, \cdot \}_2$, is defined by

$$\begin{aligned} [\{ \cdot, \cdot \}_1, \{ \cdot, \cdot \}_2]_{\mathrm{SN}}\,(F, G, H) \;&:=\; \{\{F, G\}_1, H\}_2 + \circlearrowleft (F, G, H) \\ &+\; \{\{F, G\}_2, H\}_1 + \circlearrowleft (F, G, H), \end{aligned}$$

for all $F, G, H \in \mathscr{F}(\mathcal{M})$.

Proof. It is enough to notice that condition (2.3) is nothing but the Jacobi identity for $\{ \cdot, \cdot \}$:

$$[\{ \cdot, \cdot \}, \{ \cdot, \cdot \}]_{\mathrm{SN}}\,(F, G, H) = 2\left(\{\{F, G\}, H\} + \circlearrowleft (F, G, H)\right).$$

- The Jacobi identity for $\{ \cdot, \cdot \}$ has some fundamental consequences:

(a) The set of all integrals of motion of Φ_t is an Abelian subalgebra of $\mathscr{F}(\mathcal{M})$ ("Poisson Theorem").

(b) The map

$$\mathscr{F}(\mathcal{M}) \to \mathsf{Ham}(\mathcal{M}) : F \mapsto \mathsf{v}_F$$

defines a Lie algebra anti-homomorphism:

$$[\,\mathsf{v}_F, \mathsf{v}_G\,] = -\mathsf{v}_{\{ F, G \}} \qquad \forall\, F, G \in \mathscr{F}(\mathcal{M}). \tag{2.4}$$

Proof. Let $H \in \mathscr{F}(\mathcal{M})$. Then we have:

$$\begin{aligned} [\,\mathsf{v}_F, \mathsf{v}_G\,]\,[H] &= \mathsf{v}_F[\mathsf{v}_G[H]] - \mathsf{v}_G[\mathsf{v}_F[H]] \\ &= \mathsf{v}_F[\{ H, G \}] - \mathsf{v}_G[\{ H, F \}] \\ &= \{\{ H, G \}, F \} - \{\{ H, F \}, G \} \\ &= -\{H, \{ F, G \}\} \\ &= -\mathsf{v}_{\{ F, G \}}[H]. \end{aligned}$$

(c) As a consequence of (b) and of the fact that two flows commute if and only if their vector fields have vanishing Lie bracket, we find that two local Hamiltonian flows, whose Hamiltonian vector fields are $\mathsf{v}_F := \{\,\cdot\,, F\,\}$ and $\mathsf{v}_G := \{\,\cdot\,, G\,\}$, commute if and only if the Poisson bracket $\{\,F, G\,\}$ is locally constant (not necessarily zero).

(d) Any Hamiltonian vector field $\mathsf{v}_H \in \mathsf{Ham}(\mathcal{M})$ leaves the Poisson bracket $\{\,\cdot,\cdot\,\}$ invariant. Equivalently, if P is the Poisson 2-vector field associated with the Poisson structure $\{\,\cdot,\cdot\,\}$ then $\mathfrak{L}_{\mathsf{v}_H}\mathsf{P} = 0$.

Proof. Identifying $\mathsf{P}[F,G] \equiv \mathsf{P}(\mathsf{d}F, \mathsf{d}G)$ and using (1.24) we find that for any $F, G \in \mathscr{F}(\mathcal{M})$ there holds

$$\begin{aligned}
\mathfrak{L}_{\mathsf{v}_H}\mathsf{P}[F,G] &= \mathsf{v}_H[\mathsf{P}[F,G]] - \mathsf{P}[\mathsf{v}_H[F],G] - \mathsf{P}[F,\mathsf{v}_H[G]] \\
&= \mathsf{v}_H[\{F,G\}] - \mathsf{P}[\{F,H\},G] - \mathsf{P}[F,\{G,H\}] \\
&= \{\{F,G\},H\} - \{\{F,H\},G\} - \{F,\{G,H\}\} \\
&= 0,
\end{aligned}$$

thanks to the Jacobi identity.

More generally, any vector field v on $\mathcal{M}$ which leaves $\{\,\cdot,\cdot\,\}$ invariant, i.e., $\mathfrak{L}_{\mathsf{v}}\mathsf{P} = 0$, is called *Poisson vector field*. Of course, any Hamiltonian vector field is a Poisson vector field, but the converse is not true in general.

- An equivalent way to claim that any $\mathsf{v}_H \in \mathsf{Ham}(\mathcal{M})$ leaves $\{\,\cdot,\cdot\,\}$ invariant is to say that any Hamiltonian flow Φ_t is a *Poisson morphism*. In formulas,

$$\Phi_t^\star\{\,F,G\,\} = \{\,\Phi_t^\star F, \Phi_t^\star G\,\} \qquad \forall\, F,G \in \mathscr{F}(\mathcal{M}).$$

(a) Recall that if $\mathcal{M}_1$ and $\mathcal{M}_2$ are manifolds then a smooth map $\varphi : \mathcal{M}_1 \to \mathcal{M}_2$ is a *morphism* if and only if $\varphi^\star F \in \mathscr{F}(\mathcal{M}_1)$ for all $F \in \mathscr{F}(\mathcal{M}_2)$.

(b) Let $(\mathcal{M}_1, \{\,\cdot,\cdot\,\}_1)$ and $(\mathcal{M}_2, \{\,\cdot,\cdot\,\}_2)$ be two Poisson manifolds. In general, a morphism $\varphi : \mathcal{M}_1 \to \mathcal{M}_2$ is a *Poisson morphism* if

$$\varphi^\star\{F,G\}_2 = \{\varphi^\star F, \varphi^\star G\}_1 \qquad \forall\, F,G \in \mathscr{F}(\mathcal{M}_2).$$

The inverse of φ, if exists, is also a Poisson morphism. One says in this case that φ is a *Poisson isomorphism*.

(c) An immersed submanifold $\mathcal{M}'$ of $\mathcal{M}$ is called an immersed *Poisson submanifold* if it admits a Poisson structure for which the inclusion map $\mathcal{M}' \hookrightarrow \mathcal{M}$ is a Poisson morphism. One can prove that a Poisson structure on an immersed Poisson submanifold $\mathcal{M}'$, if it exists, is unique.

- Poisson vector fields appear naturally in the context of group actions.

(a) Let $\mathbf{G}$ be a Lie group which acts on a Poisson manifold $(\mathcal{M}, \{\,\cdot,\cdot\,\})$ by means of the action $\varphi : \mathbf{G} \times \mathcal{M} \to \mathcal{M}$. Let $\mathsf{X} \in \mathfrak{g}$ and suppose that for $|t|$

sufficiently small one has that $\varphi_h : \mathcal{M} \to \mathcal{M}, h = \exp(t\mathsf{X})$, is a morphism of Poisson manifolds. Then the fundamental vector field $\underline{\mathsf{X}}$, corresponding to X,

$$\underline{\mathsf{X}}[F] := \left.\frac{\mathsf{d}}{\mathsf{d}t}\right|_{t=0} \left(\varphi^{\star}_{\exp(t\mathsf{X})} F\right) \qquad \forall F \in \mathscr{F}(\mathcal{M})$$

is a Poisson vector field. As a consequence, if for every $h \in \mathbf{G}$ the map $\varphi_h : \mathcal{M} \to \mathcal{M}$ is a Poisson morphism then all fundamental vector fields of the action are Poisson vector fields.

Proof. To show that $\mathfrak{L}_{\underline{\mathsf{X}}}\mathsf{P} = 0$ we need to check, according to (1.24), that the Leibniz property

$$\underline{\mathsf{X}}[\{F,G\}] = \{\underline{\mathsf{X}}[F],G\} + \{F,\underline{\mathsf{X}}[G]\}$$

holds for any $F,G \in \mathscr{F}(\mathcal{M})$. Now, the fact that $\varphi_{\exp(t\mathsf{X})}$ is Poisson means that

$$\varphi^{\star}_{\exp(t\mathsf{X})}\{F,G\} = \left\{\varphi^{\star}_{\exp(t\mathsf{X})}F, \varphi^{\star}_{\exp(t\mathsf{X})}G\right\}.$$

Therefore,

$$\begin{aligned}
\underline{\mathsf{X}}[\{F,G\}] &= \left.\frac{\mathsf{d}}{\mathsf{d}t}\right|_{t=0} \varphi^{\star}_{\exp(t\mathsf{X})}\{F,G\} \\
&= \left.\frac{\mathsf{d}}{\mathsf{d}t}\right|_{t=0} \left\{\varphi^{\star}_{\exp(t\mathsf{X})}F, \varphi^{\star}_{\exp(t\mathsf{X})}G\right\} \\
&= \left\{\left.\frac{\mathsf{d}}{\mathsf{d}t}\right|_{t=0}\left(\varphi^{\star}_{\exp(t\mathsf{X})}F\right), G\right\} + \left\{F, \left.\frac{\mathsf{d}}{\mathsf{d}t}\right|_{t=0}\left(\varphi^{\star}_{\exp(t\mathsf{X})}G\right)\right\} \\
&= \{\underline{\mathsf{X}}[F],G\} + \{F,\underline{\mathsf{X}}[G]\},
\end{aligned}$$

as desired.

(b) If a fundamental vector field is also Hamiltonian then φ is called *Hamiltonian action*. In such a case one can build a map $\mu : \mathcal{M} \to \mathfrak{g}^*$ by defining for any $p \in \mathcal{M}$ and for any $\mathsf{X} \in \mathfrak{g}$ a smooth function

$$F_{\mathsf{X}}(p) := \langle\, \mu(p), \mathsf{X}\,\rangle\,.$$

Here F_{X} is a Hamiltonian function for X (defined up to an additive Casimir function). The map μ is called *momentum map*.

(c) The *co-momentum map* is the map dual to the momentum map. It is a linear map $F : \mathfrak{g} \to \mathscr{F}(\mathcal{M})$ having the property that for any $\mathsf{X} \in \mathfrak{g}$ the function F_{X} is a Hamiltonian for the fundamental vector field $\underline{\mathsf{X}}$, i.e.,

$$\underline{\mathsf{X}} = \mathsf{v}_{F_{\mathsf{X}}} = \{\,\cdot, F_{\mathsf{X}}\,\}.$$

(d) The famous "Noether Theorem", in its Hamiltonian form, states that if the action φ is Hamiltonian, with co-momentum map $F : \mathfrak{g} \to \mathscr{F}(\mathcal{M})$, and if $H \in \mathscr{F}(\mathcal{M})$ is a $\mathbf{G}$-invariant function, then for any $\mathsf{X} \in \mathfrak{g}$ the function F_{X} is an integral of motion for v_H.

Proof. In general, $\mathbf{G}$-invariance of H means $H(\varphi(h,p)) = H(p)$ for all $h \in \mathbf{G}$ and $p \in \mathcal{M}$, so that

$$\begin{aligned} -\mathsf{v}_H(p)[F_{\mathsf{X}}] &= \{H, F_{\mathsf{X}}\}(p) = \mathsf{v}_{F_{\mathsf{X}}}(p)[H] = \underline{\mathsf{X}}(p)[H] \\ &= \left.\frac{\mathsf{d}}{\mathsf{d}t}\right|_{t=0} H\left(\varphi_{\exp(t\mathsf{X})}\, p\right) = \left.\frac{\mathsf{d}}{\mathsf{d}t}\right|_{t=0} H(p) = 0, \end{aligned}$$

which shows that F_{X} is an integral of motion for v_H.

- It may happen that a Poisson manifold can be equipped with several independent Poisson structures. Such circumstance is quite remarkable. Indeed, we will prove in the next Chapter that if a dynamical system on $\mathcal{M}$ admits more independent Hamiltonian formulations, w.r.t. distinct compatible Poisson brackets, then it automatically possesses a large number of integrals of motion. Such systems are called *multi-Hamiltonian systems.*

(a) Let $s \in \mathbb{N}$, $s > 1$. Then $(\mathcal{M}, \{\cdot,\cdot\}_1, \dots, \{\cdot,\cdot\}_s)$ is called *multi-Hamiltonian manifold* if each $\{\cdot,\cdot\}_i$, $i = 1, \dots, s$, is a Poisson structure on $\mathcal{M}$ and if any linear combination of them is also a Poisson structure. In such a case the s Poisson structures are called *compatible.* A vector field which is Hamiltonian w.r.t. all Poisson structures on $\mathcal{M}$ is called *multi-Hamiltonian vector field.*

(b) If $\{\cdot,\cdot\}_1$ and $\{\cdot,\cdot\}_2$ are two Poisson structures then the bracket defined by $\lambda_1\{\cdot,\cdot\}_1 + \lambda_2\{\cdot,\cdot\}_2$, for non-zero scalars λ_1, λ_2, is a Poisson bracket if and only if

$$[\lambda_1\{\cdot,\cdot\}_1 + \lambda_2\{\cdot,\cdot\}_2, \lambda_1\{\cdot,\cdot\}_1 + \lambda_2\{\cdot,\cdot\}_2]_{\mathrm{SN}} = 0,$$

that is $[\{\cdot,\cdot\}_1, \{\cdot,\cdot\}_2]_{\mathrm{SN}} = 0$. In such a case $\{\cdot,\cdot\}_1$ and $\{\cdot,\cdot\}_2$ are compatible. One say that $\lambda_1\{\cdot,\cdot\}_1 + \lambda_2\{\cdot,\cdot\}_2$ is a *Poisson pencil.* Similarly, s Poisson structures $\{\cdot,\cdot\}_1, \dots, \{\cdot,\cdot\}_s$ are compatible if and only if they are pairwise compatible.

(c) Let $(\mathcal{M}, \{\cdot,\cdot\}_1)$ be a Poisson manifold and v be a vector field on $\mathcal{M}$. Denote by P_1 the Poisson 2-vector field associated with $\{\cdot,\cdot\}_1$. If $\mathsf{P}_2 := \mathfrak{L}_{\mathsf{v}}\mathsf{P}_1 \in \mathfrak{X}^2(\mathcal{M})$ satisfies the Jacobi identity then $\{\cdot,\cdot\}_1$ and $\{\cdot,\cdot\}_2$, induced by P_2, are compatible Poisson structures.

Proof. $\{\cdot,\cdot\}_1$ and $\{\cdot,\cdot\}_2$ are compatible if and only if their sum $\{\cdot,\cdot\}_1 + \{\cdot,\cdot\}_2$ satisfies the Jacobi identity. Since $\{\cdot,\cdot\}_1$ and $\{\cdot,\cdot\}_2$ satisfy the Jacobi identity this means that we need to check that

$$\{\{F,G\}_1, H\}_2 + \{\{F,G\}_2, H\}_1 + \circlearrowleft (F,G,H) = 0. \tag{2.5}$$

Using formula (1.24), we see that

$$\mathsf{P}_2\,[F,G] := (\mathfrak{L}_{\mathsf{v}}\mathsf{P}_1)\,[F,G] = \mathsf{v}\,[\mathsf{P}_1\,[F,G]] - \mathsf{P}_1\,[\mathsf{v}[F],G] - \mathsf{P}_1\,[F,\mathsf{v}[G]]\,.$$

Identifying $\mathsf{P}_i[F,G] \equiv \mathsf{P}_i(\mathrm{d}F,\mathrm{d}G)$, $i = 1,2$, we get

$$\{F,G\}_2 = \mathsf{v}[\{F,G\}_1] - \{\mathsf{v}[F],G\}_1 - \{F,\mathsf{v}[G]\}_1,$$

so that we can express all brackets $\{\,\cdot,\cdot\,\}_2$ in (2.5) in terms of $\{\,\cdot,\cdot\,\}_1$. The equality (2.5) is then a consequence of the Jacobi identity for $\{\,\cdot,\cdot\,\}_1$.

► We now give some coordinate-dependent formulas. Let us consider a coordinate chart $\mathcal{U}$ of $\mathcal{M}$ where a point $p \in \mathcal{M}$ can be expressed by local coordinates $x := (x_1,\ldots,x_n) : \mathcal{U} \to \mathbb{R}^n$.

- A Poisson structure on $\mathcal{U}$ is locally defined in terms of its *structure functions*

$$x_{ij} := \{\,x_i,x_j\,\} \in \mathscr{F}(\mathcal{U}), \qquad i,j = 1,\ldots,n,$$

which define a skew-symmetric matrix

$$\mathcal{X} := \left(x_{ij}\right)_{1\leqslant i,j\leqslant n}, \tag{2.6}$$

called *Poisson matrix* associated with $\{\,\cdot,\cdot\,\}$. The structure functions satisfy the equations

$$x_{ij} = -x_{ji}, \qquad \forall\, i,j = 1,\ldots,n, \tag{2.7}$$

and

$$\sum_{\ell=1}^{n}\left(\frac{\partial x_{ij}}{\partial x_\ell}x_{\ell k} + \frac{\partial x_{jk}}{\partial x_\ell}x_{\ell i} + \frac{\partial x_{ki}}{\partial x_\ell}x_{\ell j}\right) = 0, \qquad \forall\, i,j,k = 1,\ldots,n. \tag{2.8}$$

Formula (2.8) is the coordinate representation of the Jacobi identity. Note that the skew-symmetry of $\mathcal{X}$ implies that its rank is always an even number.

- Conversely, every set of functions $x_{ij} \in \mathscr{F}(\mathcal{U})$ satisfying conditions (2.7-2.8) locally defines a Poisson bracket on $\mathcal{M}$ between functions $F,H \in \mathscr{F}(\mathcal{U})$ by setting

$$\{\,F(x),H(x)\,\} := \sum_{1\leqslant i<j\leqslant n} x_{ij}\left(\frac{\partial F}{\partial x_i}\frac{\partial H}{\partial x_j} - \frac{\partial H}{\partial x_i}\frac{\partial F}{\partial x_j}\right) = [\mathrm{d}F]^\top \mathcal{X}\,[\mathrm{d}H], \tag{2.9}$$

where $[\mathrm{d}F]$ is the column vector which represents $\mathrm{d}F$ in the basis $\{\mathrm{d}x_1,\ldots,\mathrm{d}x_n\}$, i.e., the i-th component of $[\mathrm{d}F]$ is $\partial F/\partial x_i$.

- For $H \in \mathscr{F}(\mathcal{U})$ the Hamiltonian vector field v_H, which is the infinitesimal generator of a Hamiltonian flow Φ_t, can be expressed as

$$\mathsf{v}_H = \sum_{j=1}^{n}\{\,x_j,H(x)\,\}\frac{\partial}{\partial x_j}.$$

- The local form of *Hamilton equations* expressing the time evolution of coordinates $x := (x_1, \ldots, x_n)$, is given by

$$\mathsf{v}_{x_i} \equiv \mathsf{v}_H [x_i] \equiv \dot{x}_i = \{ x_i, H(x) \} = \sum_{j=1}^{n} x_{ij} \frac{\partial H}{\partial x_j}, \qquad i = 1, \ldots, n. \tag{2.10}$$

- If $F \in \mathscr{F}(\mathcal{U})$ we obtain

$$\begin{aligned} \mathsf{v}_H[F(x)] &= (\mathfrak{L}_{\mathsf{v}_H} F)(x) = \sum_{i=1}^{n} \{ x_i, H(x) \} \frac{\partial F}{\partial x_i} \\ &= \sum_{1 \leqslant i < j \leqslant n} x_{ij} \left(\frac{\partial F}{\partial x_i} \frac{\partial H}{\partial x_j} - \frac{\partial H}{\partial x_i} \frac{\partial F}{\partial x_j} \right) = \{ F(x), H(x) \}. \end{aligned}$$

- In terms of the structure functions x_{ij} the Poisson 2-vector field P can be written as

$$\mathsf{P} = \sum_{1 \leqslant i < j \leqslant n} x_{ij} \frac{\partial}{\partial x_i} \wedge \frac{\partial}{\partial x_j}. \tag{2.11}$$

2.1.1 *Rank of a Poisson manifold*

► Given a Poisson manifold $(\mathcal{M}, \{ \cdot, \cdot \})$ the existence of Casimirs is related to degeneracy of the Poisson structure. Thus, we need to give a proper definition of the rank of the Poisson structure. To do so we first introduce a particular distribution on $\mathcal{M}$. If we specialize the construction of Hamiltonian vector fields to a point $p \in \mathcal{M}$ we find a linear space $\mathsf{Ham}_p(\mathcal{M}) := \{ \mathsf{v}_H(p) \ : \ H \in \mathscr{F}(\mathcal{M}) \} \subseteq T_p\mathcal{M}$, whose dimension varies in general with p, thereby defining a distribution on $\mathcal{M}$. Notice that $\mathsf{Ham}_p(\mathcal{M}) = \mathsf{span}\, \{ \mathsf{v}_{x_1}(p), \ldots, \mathsf{v}_{x_n}(p) \}$, where $(x_1, \ldots, x_n)$ is any system of local coordinates, defined on a coordinate chart containing p.

► We now give the following definition.

Definition 2.3

1. *For $p \in \mathcal{M}$, the dimension of $\mathsf{Ham}_p(\mathcal{M})$ is called the* **rank of** $\{ \cdot, \cdot \}$ **at** *p, denoted by $\mathsf{rank}_p\{ \cdot, \cdot \}$. The* **rank of the Poisson manifold** *$(\mathcal{M}, \{ \cdot, \cdot \})$ is*

$$\mathsf{rank}\{ \cdot, \cdot \} := \max_{p \in \mathcal{M}} \left(\mathsf{rank}_p\{ \cdot, \cdot \} \right).$$

2. *$(\mathcal{M}, \{ \cdot, \cdot \})$ is a* **regular Poisson manifold** *if $\mathsf{rank}_p\{ \cdot, \cdot \} = \mathsf{rank}\{ \cdot, \cdot \}$ for all $p \in \mathcal{M}$.*
3. *The Poisson structure $\{ \cdot, \cdot \}$ has* **maximal rank at** *$p \in \mathcal{M}$ if $\mathsf{rank}_p\{ \cdot, \cdot \} = n$.*

The Poisson structure $\{\cdot,\cdot\}$ *has* **maximal rank on** $\mathcal{M}$ *if* $\mathsf{rank}_p\{\cdot,\cdot\} = n$ *for all* $p \in \mathcal{M}$, *i.e.,* $\mathsf{Ham}_p(\mathcal{M}) = T_p\mathcal{M}$ *for all* $p \in \mathcal{M}$.

► In a coordinate chart $\mathcal{U}$ of $\mathcal{M}$ where a point $p \in \mathcal{M}$ can be expressed in local coordinates $x := (x_1, \dots, x_n)$ we have

$$\mathsf{rank}_p\{\cdot,\cdot\} = \mathsf{rank}\,(\mathcal{X}(p)), \tag{2.12}$$

where $\mathcal{X}(p)$ is the Poisson matrix (2.6) at p w.r.t. the chart $\mathcal{U}$.

- The skew-symmetry of $\mathcal{X}$ implies that the rank of a Poisson structure at a point is always an even number.
- The rank (2.12) does not depend on the choice of coordinates on $\mathcal{U}$.
- If a Poisson manifold has maximal rank on $\mathcal{M}$ then its dimension n is necessarily even.

► We now give two claims (without proof).

1. Let $s \in \mathbb{N}$ with $2s \leqslant \mathsf{rank}\{\cdot,\cdot\} \leqslant n$. Then the subset of $\mathcal{M}$ defined by

   $$\mathcal{M}_{(s)} := \{p \in \mathcal{M} \,:\, \mathsf{rank}_p\{\cdot,\cdot\} \geqslant 2s\},$$

   is open. Note that:

   (a) In view of the fact that any Hamiltonian vector field $\mathsf{v} \in \mathsf{Ham}(\mathcal{M})$ leaves $\{\cdot,\cdot\}$ invariant there follows that $\mathcal{M}_{(s)}$ is an invariant set for all Hamiltonian flows.

   (b) If we denote the rank of $\{\cdot,\cdot\}$ by $2r$, then $\mathcal{M}_{(r)}$ is a non-empty open subset of $\mathcal{M}$ and the restriction of $\{\cdot,\cdot\}$ to $\mathcal{M}_{(r)}$ is regular (of rank $2r$).

2. Let $(\mathcal{M}_i, \{\cdot,\cdot\}_i)$, $i = 1, 2$, be two Poisson manifolds. Let $p \in \mathcal{M}_1$ and $\varphi : \mathcal{M}_1 \to \mathcal{M}_2$ be a Poisson morphism. Then,

   $$\mathsf{rank}_p\{\cdot,\cdot\}_1 \geqslant \mathsf{rank}_{\varphi(p)}\{\cdot,\cdot\}_2.$$

   This statement means that the inequality

   $$\dim\left(\mathsf{Ham}_p(\mathcal{M})\right) \leqslant \dim \mathcal{M} = n$$

   implies that a necessary condition for a submanifold $\mathcal{M}'$ of $\mathcal{M}$ to be a Poisson submanifold is that

   $$\dim \mathcal{M}' \geqslant \mathsf{rank}_p\{\cdot,\cdot\}$$

   for all $p \in \mathcal{M}'$.

► The next definition will allow us to obtain some important results about the rank of a Poisson manifold.

Definition 2.4

Let $\mathtt{F} := (F_1, \dots, F_s)$, $F_i \in \mathscr{F}(\mathcal{M})$, $i = 1, \dots, s$, *be a set of smooth functions on* $(\mathcal{M}, \{\cdot,\cdot\})$.

1. $\mathtt{F}$ *is* **Poisson involutive** *if any two elements of* $\mathtt{F}$ *are in Poisson involution, i.e.,* $\{F_i, F_j\} = 0$ *for all* $i, j = 1, \dots, s$.
2. $\mathtt{F}$ *is* **independent** *if the open subset on which the differentials* $\mathrm{d}F_1, \dots, \mathrm{d}F_s$ *are independent is dense in* $\mathcal{M}$, *i.e.,*

$$\mathcal{U}_{\mathtt{F}} := \{p \in \mathcal{M} \,:\, \mathrm{d}F_1(p) \wedge \cdots \wedge \mathrm{d}F_s(p) \neq 0\}$$

is a dense open subset of $\mathcal{M}$.

► Remarks:

- $\mathtt{F}$ defines a map $\mathcal{M} \to \mathbb{R}^s$ which we denote by the same letter.
  (a) A *fiber* of $\mathtt{F}$ is a fiber of $\mathtt{F}$ as a map: it is a common level set of the s functions F_i. We will see in the next Chapter that a particularly important instance occurs if $s = n - r$, where $2r$ is the rank of $\{\cdot,\cdot\}$.
  (b) The fiber of $\mathtt{F}$ that passes through $p \in \mathcal{M}$ will be denoted by $\mathcal{F}_p$:

$$\mathcal{F}_p := \{\tilde{p} \in \mathcal{M} \,:\, F_i(\tilde{p}) = F_i(p),\ i = 1, \dots, s\}.$$

  (c) For $c \in \mathbb{R}^s$ we denote by $\mathcal{F}_c$ the fiber $\mathtt{F}^{-1}(c)$ over c. By the "Inverse function Theorem", the fiber $\mathcal{F}_c$ over each regular value c that lies in the image of $\mathtt{F}$ is non-singular.
- Let $\mathtt{F}$ be Poisson involutive and independent. Then we have two immediate consequences:
  (a) The Hamiltonian vector fields v_{F_i}, $i = 1, \dots, s$, commute and for any $p \in \mathcal{M}$ they are tangent to the fibers of $\mathtt{F}$. This is a plane consequence of the fact that

$$\left[\mathsf{v}_{F_i}, \mathsf{v}_{F_j}\right] = -\mathsf{v}_{\{F_i, F_j\}} = 0.$$

  (b) The Leibniz property of the Poisson bracket implies that the subalgebra of $\mathscr{F}(\mathcal{M})$ generated by the functions F_i is also Poisson involutive.
- Let $x := (x_1, \dots, x_n)$ be local coordinates around $p \in \mathcal{M}$. Then $p \in \mathcal{U}_{\mathtt{F}}$ if and only if

$$\mathsf{rank}\left(\frac{\partial F_i}{\partial x_j}(p)\right)_{1 \leqslant i \leqslant s, 1 \leqslant j \leqslant n} = s.$$

In such a case one also says that the functions $F_1, \ldots, F_s$ are *functionally independent* on $\mathcal{U}_{\mathtt{F}}$. One obviously has that $s \leqslant n$.

► The number of independent functions on a smooth manifold $\mathcal{M}$ is bounded by n, the dimension of $\mathcal{M}$. In the case of a Poisson manifold, this upper bound is much lower, as given by the next claim.

Theorem 2.1

Assume that $\mathsf{rank}\{\cdot,\cdot\} = 2r \leqslant n$ *and let* $\mathtt{F} := (F_1, \ldots, F_s)$, $F_i \in \mathscr{F}(\mathcal{M})$, $i = 1, \ldots, s$, *be independent.*

1. *If* $\mathtt{F}$ *consists of only Casimirs then* $s \leqslant n - 2r$,
2. *If* $\mathtt{F}$ *is Poisson involutive then* $s \leqslant n - r$,
3. *If* $\mathtt{F}$ *is Poisson involutive with* $s = n - r$ *then*

$$\mathsf{dim}\left(\mathsf{span}\{\mathsf{v}_{F_1}(p), \ldots, \mathsf{v}_{F_s}(p)\}\right) \leqslant r$$

for any $p \in \mathcal{U}_{\mathtt{F}}$. *Equality holds if* $p \in \mathcal{U}_{\mathtt{F}} \cap \mathcal{M}_{(r)}$.

Proof. Let P be the Poisson 2-vector field associated with $\{\cdot,\cdot\}$ and consider, for $p \in \mathcal{M}$, the map $\widetilde{\mathsf{P}}_p : T_p^*\mathcal{M} \to T_p\mathcal{M}$, simply by setting

$$\widetilde{\mathsf{P}}_p(\mathsf{d}F(p)) := \mathsf{P}(\cdot, \mathsf{d}F(p)) = \{\cdot, F\}(p) = \mathsf{v}_F(p) \qquad \forall F \in \mathscr{F}(\mathcal{M}).$$

The rank of $\widetilde{\mathsf{P}}_p$ is $\mathsf{rank}_p\{\cdot,\cdot\}$. Let $p \in \mathcal{U}_{\mathtt{F}} \cap \mathcal{M}_{(r)}$. We prove all claims.

1. Suppose that $\mathtt{F}$ consists of only Casimirs and recall that if $F \in \mathsf{Cas}(\mathcal{M})$ then $\mathsf{d}F(p) \in \mathsf{Ker}\,\widetilde{\mathsf{P}}_p$, whose dimension is $n - \mathsf{rank}_p\{\cdot,\cdot\}$. Since $\mathsf{d}F_1(p), \ldots, \mathsf{d}F_s(p)$ are independent we have that

$$s \leqslant \mathsf{dim}\left(\mathsf{Ker}\,\widetilde{\mathsf{P}}_p\right) = n - 2r,$$

which is the first claim.

2. Suppose that $\mathtt{F}$ is Poisson involutive and consider the fiber $\mathcal{F}_p$. The restriction of $\mathcal{F}_p$ to a neighborhood $\mathcal{U}$ of p is a submanifold of $\mathcal{U}$ of dimension $n - s$, passing through p. This dimension is an upper bound for the dimension d_p of $\mathsf{span}\{\mathsf{v}_{F_1}(p), \ldots, \mathsf{v}_{F_s}(p)\}$, because these s vectors are tangent to that fiber at p, that is

$$d_p \leqslant n - s.$$

Moreover the differentials $\mathsf{d}F_1(p), \ldots, \mathsf{d}F_s(p)$ are independent at p, so that

$$d_p \geqslant s - \mathsf{dim}\left(\mathsf{Ker}\,\widetilde{\mathsf{P}}_p\right) = s + 2r - n.$$

Combining the two inequalities for d_p we get

$$s + 2r - n \leqslant d_p \leqslant n - s, \tag{2.13}$$

that is $s \leqslant n - r$ as desired.

3. Suppose that $\mathtt{F}$ is Poisson involutive and $s = n - r$. From (2.13) we obtain $r \leqslant d_p \leqslant r$, so that

$$d_p := \mathsf{dim}\left(\mathsf{span}\{\mathsf{v}_{F_1}(p), \ldots, \mathsf{v}_{F_s}(p)\}\right) = r.$$

The Theorem is proved. ■

► The extreme case $s = n - r$ is of particular importance. In this case, the commuting Hamiltonian vector fields $\mathsf{v}_{F_1}, \ldots, \mathsf{v}_{F_{n-r}}$ define an integrable (in the sense of Frobenius) distribution of rank r on $\mathcal{U}_{\mathtt{F}} \cap \mathcal{M}_{(r)}$ (see Lemma 3.1).

2.1.2 *Examples*

► Let us present some examples of Poisson structures.

Example 2.1 (*The constant Poisson structure*)

Any constant skew-symmetric $n \times n$ matrix is the matrix of a Poisson structure on $\mathbb{R}^n$, in terms of its standard coordinates, as follows from (2.8). By the classification theorem for skew-symmetric bilinear forms there exists a linear system of coordinates $(x_1, \ldots, x_n)$ of $\mathbb{R}^n$ w.r.t. which the Poisson matrix takes the form

$$\mathcal{X} := \begin{pmatrix} \mathsf{0}_r & \mathsf{1}_r & \mathsf{0}_{r\times(n-2r)} \\ -\mathsf{1}_r & \mathsf{0}_r & \mathsf{0}_{r\times(n-2r)} \\ \mathsf{0}_{(n-2r)\times r} & \mathsf{0}_{(n-2r)\times r} & \mathsf{0}_{(n-2r)\times(n-2r)} \end{pmatrix}.$$

The integer $2r$ is the rank of the Poisson structure.

Example 2.2 (*The canonical Hamiltonian phase space* $\mathbb{R}^{2n}$)

Any constant skew-symmetric matrix on $\mathbb{R}^{2n}$ is the matrix of a Poisson structure in terms of its coordinates $x := (x_1, \ldots, x_{2n})$, as follows from (2.8).

- The non-degenerate *canonical Poisson structure* on $\mathbb{R}^{2n}$ defined by

$$\mathcal{X}_{\text{can}} := \begin{pmatrix} \mathsf{0}_n & \mathsf{1}_n \\ -\mathsf{1}_n & \mathsf{0}_n \end{pmatrix}, \tag{2.14}$$

is an example of constant Poisson structure on $\mathbb{R}^{2n}$ with maximal rank $2n$.

- Defining coordinates $(x_1, \ldots, x_{2n}) := (q_1, \ldots, q_n, p_1, \ldots, p_n)$ we can write the Poisson 2-vector field corresponding to (2.14) as

$$\mathsf{P}_{\text{can}} = \sum_{i=1}^{n} \frac{\partial}{\partial q_i} \wedge \frac{\partial}{\partial p_i},$$

which implies the following canonical Poisson brackets

$$\{p_i, p_j\} = \{q_i, q_j\} = 0, \qquad \{q_i, p_j\} = \delta_{ij}, \qquad i, j = 1, \ldots, n.$$

- If $H \in \mathscr{F}(\mathbb{R}^{2n})$ then, by using (2.2), we can construct the corresponding canonical Hamiltonian vector field by contracting the Poisson 2-vector P with the differential of H,

$$\mathrm{d}H = \sum_{i=1}^{n} \left(\frac{\partial H}{\partial q_i} \mathrm{d}q_i + \frac{\partial H}{\partial p_i} \mathrm{d}p_i \right).$$

From formula (2.2) we obtain:

$$\mathsf{v}_H = \widetilde{\mathsf{P}}_{\mathrm{can}}(\mathrm{d}H) = \mathsf{P}_{\mathrm{can}}(\cdot, \mathrm{d}H) = -\mathsf{P}_{\mathrm{can}}(\mathrm{d}H, \cdot) = \sum_{i=1}^{n} \left(\frac{\partial H}{\partial p_i} \frac{\partial}{\partial q_i} - \frac{\partial H}{\partial q_i} \frac{\partial}{\partial p_i} \right),$$

which is nothing but that the canonical Hamiltonian vector field on $\mathbb{R}^{2n}$.

Example 2.3 (*A Poisson structure on* $\mathbb{R}^3$)

Define the following 3×3 real skew-symmetric matrix:

$$\mathcal{X} := \begin{pmatrix} 0 & f_3(x) & -f_2(x) \\ -f_3(x) & 0 & f_1(x) \\ f_2(x) & -f_1(x) & 0 \end{pmatrix},$$

where $f_i \in \mathscr{F}(\mathbb{R}^3)$, $i = 1, 2, 3$, and $x := (x_1, x_2, x_3)^\top$.

- $\mathcal{X}$ defines a Poisson matrix of a Poisson structure on $\mathbb{R}^3$ if and only if there holds

$$\langle \mathsf{curl}_x f(x), f(x) \rangle = 0 \qquad \forall\, x \in \mathbb{R}^3. \tag{2.15}$$

Here $f(x) := (f_1(x), f_2(x), f_3(x))^\top$ and

$$\mathsf{curl}_x f(x) = \left(\frac{\partial f_3}{\partial x_2} - \frac{\partial f_2}{\partial x_3}, \frac{\partial f_1}{\partial x_3} - \frac{\partial f_3}{\partial x_1}, \frac{\partial f_2}{\partial x_1} - \frac{\partial f_1}{\partial x_2} \right)^\top.$$

Note that (2.15) is a set of three scalar partial differential equations corresponding to the Jacobi identity for $\mathcal{X}$.

- For $F_1 = F_1(x)$ and $F_2 = F_2(x)$ arbitrary polynomials in $\mathbb{R}^3$, the function defined by

$$f(x) := F_1(x)\ \mathsf{grad}_x F_2(x)$$

is a particular solution of (2.15). Such a function leads to a large number of regular Poisson structures on $\mathbb{R}^3$. Explicitly one finds

$$\mathcal{X} = F_1(x) \begin{pmatrix} 0 & \dfrac{\partial F_2}{\partial x_3} & -\dfrac{\partial F_2}{\partial x_2} \\ -\dfrac{\partial F_2}{\partial x_3} & 0 & \dfrac{\partial F_2}{\partial x_1} \\ \dfrac{\partial F_2}{\partial x_2} & -\dfrac{\partial F_2}{\partial x_1} & 0 \end{pmatrix}.$$

In coordinates we get the following Poisson brackets:

$$\{x_1, x_2\} = F_1(x) \frac{\partial F_2}{\partial x_3}, \quad \{x_2, x_3\} = F_1(x) \frac{\partial F_2}{\partial x_1}, \quad \{x_3, x_1\} = F_1(x) \frac{\partial F_2}{\partial x_2}.$$

- Generically, the matrix $\mathcal{X}$ has rank two, while the dimension of the phase space is three. One can verify that the function F_2 is a Casimir function of the Poisson structure: the matrix $\mathcal{X}$ annihilates the gradient of F_2:

$$\mathcal{X}\ \mathsf{grad}_x F_2(x) = 0.$$

Example 2.4 (*The Euler top*)

Consider the Poisson structure of Example 2.3 and define

$$F_1(x) := 1, \qquad F_2(x) := \frac{1}{2}\left(x_1^2 + x_2^2 + x_3^2\right).$$

- The resulting Poisson matrix is

$$\mathcal{X} = \begin{pmatrix} 0 & x_3 & -x_2 \\ -x_3 & 0 & x_1 \\ x_2 & -x_1 & 0 \end{pmatrix},$$

which corresponds to the Poisson 2-vector field (see (2.11))

$$\mathsf{P} = x_1 \frac{\partial}{\partial x_2} \wedge \frac{\partial}{\partial x_3} + x_2 \frac{\partial}{\partial x_3} \wedge \frac{\partial}{\partial x_1} + x_3 \frac{\partial}{\partial x_1} \wedge \frac{\partial}{\partial x_2}. \tag{2.16}$$

Note that $\mathcal{X}$ defines the structure constants (1.32) of the three-dimensional orthogonal Lie algebra described in Example 1.17:

$$\{ x_i, x_j \} = \varepsilon_{ijk}\, x_k. \tag{2.17}$$

- We also have

$$\{ F(x), G(x) \} = \langle x, \mathrm{grad}_x F(x) \times \mathrm{grad}_x G(x) \rangle, \qquad F, G \in \mathscr{F}(\mathbb{R}^3). \tag{2.18}$$

The bracket (2.18) is an example of a Lie-Poisson bracket. Indeed we will show that the Euler top (defined below) is a Hamiltonian system defined on the Poisson manifold $\mathfrak{so}(3)^* \simeq \mathbb{R}^3$ (see Example 1.22 where $\mathfrak{so}(3)$ has been identified with $\mathfrak{so}(3)^*$).

- Now consider the smooth function

$$H(x) := \frac{1}{2}\left(a_1\, x_1^2 + a_2\, x_2^2 + a_3\, x_3^2\right),$$

with $a_1, a_2, a_3 \in \mathbb{R}$. We can construct a Hamiltonian vector field in terms of the action of $\mathcal{X}$ on the gradient of H:

$$\mathcal{X}\ \mathrm{grad}_x H(x) = ((a_2 - a_3)\, x_2\, x_3, (a_3 - a_1)\, x_3\, x_1\, (a_1 - a_2)\, x_1\, x_2). \tag{2.19}$$

- In a more geometric language the vector field (2.19) is constructed by contracting the Poisson 2-vector field (2.16) with the differential of H,

$$\mathrm{d}H = a_1\, x_1\, \mathrm{d}x_1 + a_2\, x_2\, \mathrm{d}x_2 + a_3\, x_3\, \mathrm{d}x_3.$$

Indeed, by using (2.2) we obtain

$$\begin{aligned} \mathsf{v}_H &= \widetilde{\mathsf{P}}(\mathrm{d}H) = \mathsf{P}(\cdot, \mathrm{d}H) = -\mathsf{P}(\mathrm{d}H, \cdot) \\ &= (a_2 - a_3)\, x_2\, x_3 \frac{\partial}{\partial x_1} + (a_3 - a_1)\, x_3\, x_1 \frac{\partial}{\partial x_2} + (a_1 - a_2)\, x_1\, x_2 \frac{\partial}{\partial x_3}. \end{aligned}$$

Such a vector field defines exactly the *Euler top* (1.2):

$$\begin{cases} \dot{x}_1 = (a_2 - a_3)\, x_2\, x_3, \\ \dot{x}_2 = (a_3 - a_1)\, x_3\, x_1, \\ \dot{x}_3 = (a_1 - a_2)\, x_1\, x_2. \end{cases} \tag{2.20}$$

This is probably the most famous Hamiltonian system described in terms of an odd number of ODEs. Indeed, the configuration manifold of the three-dimensional Euler top is the Lie group $\mathbf{SO}(3)$, which is usually parametrized in terms of three *Euler angles*.

- Note that we can obtain (2.20) also by using (2.10). For instance,

$$\begin{aligned} \mathsf{v}_H\,[x_1] &\equiv \dot{x}_1 = \{\,x_1, H(x)\,\} = x_{11}\frac{\partial H}{\partial x_1} + x_{12}\frac{\partial H}{\partial x_2} + x_{13}\frac{\partial H}{\partial x_3} \\ &= (a_2 - a_3)\,x_2\,x_3. \end{aligned}$$

- The function H is the Hamiltonian of the system, an integral of motion:

$$\mathsf{v}_H[H(x)] = (\mathfrak{L}_{\mathsf{v}_H} H)\,(x) = \dot{x}_1\frac{\partial H}{\partial x_1} + \dot{x}_2\frac{\partial H}{\partial x_2} + \dot{x}_3\frac{\partial H}{\partial x_3} = 0$$

  The function F_2 is a Casimir, namely another integral of motion, functionally independent on H, which generates a trivial dynamics. Therefore the Euler top admits two functionally independent integrals of motion which are in Poisson involution w.r.t. the Poisson structure (2.17).

- It is easy to see that v_H is a bi-Hamiltonian vector field. Indeed, one can define two compatible Poisson structures associated with $\mathcal{X}_1 := \mathcal{X}$ and

$$\mathcal{X}_2 := \begin{pmatrix} 0 & a_3\,x_3 & -a_2\,x_2 \\ -a_3\,x_3 & 0 & a_1\,x_1 \\ a_2\,x_2 & -a_1\,x_1 & 0 \end{pmatrix}.$$

  It turns out that

$$\mathsf{v}_H = \{\,\cdot, F_2\,\}_2 = \{\,\cdot, H\,\}_1, \qquad \{\,\cdot, F_2\,\}_1 = \{\,\cdot, H\,\}_2 = 0.$$

Example 2.5 (*The Lie-Poisson structure*)

An important example of Poisson manifold is given by *Lie-Poisson structure* on the dual $\mathfrak{g}^*$ of a finite-dimensional Lie algebra $\mathfrak{g}$. An examples of Lie-Poisson structure has been already considered in Example 2.4. For smooth functions $F, G \in \mathscr{F}(\mathfrak{g}^*)$ on $\mathfrak{g}^*$ the *Lie-Poisson bracket* is defined by

$$\{F, G\}_{\mathrm{LP}}(\mathsf{X}^*) := \langle\,\mathsf{X}^*, [\,\mathrm{d}F(\mathsf{X}^*), \mathrm{d}G(\mathsf{X}^*)\,]\,\rangle, \qquad \mathsf{X}^* \in \mathfrak{g}^*, \tag{2.21}$$

where $\mathrm{d}F(\mathsf{X}^*)$ and $\mathrm{d}G(\mathsf{X}^*)$ are interpreted as elements of $\mathfrak{g}$ when computing the bracket. While bilinearity and skew-symmetry of (2.21) are evident, the Jacobi identity is not. One can prove that the Jacobi identity of the bracket (2.21) is a consequence of the Jacobi identity for $\mathfrak{g}$.

Example 2.6 (*A family of quadratic Poisson structure on* $\mathbb{R}^4$)

A family of quadratic Poisson structures on $\mathbb{R}^4$ is the following. Let $x := (x_1, x_2, x_3, x_4)$ be coordinates on $\mathbb{R}^4$ and define the skew-symmetric matrix

$$\mathcal{X} := \begin{pmatrix} 0 & b_1\,x_3\,x_4 & b_2\,x_2\,x_4 & b_3\,x_2\,x_3 \\ -b_1\,x_3\,x_4 & 0 & a_3\,x_1\,x_4 & a_2\,x_1\,x_3 \\ -b_2\,x_2\,x_4 & -a_3\,x_1\,x_4 & 0 & a_1\,x_1\,x_2 \\ -b_3\,x_2\,x_3 & -a_2\,x_1\,x_3 & -a_1\,x_1\,x_2 & 0 \end{pmatrix},$$

where $a_1, a_2, a_3, b_1, b_2, b_3 \in \mathbb{R}$ are parameters.

- Four checks of the Jacobi identity suffice to show that this is a Poisson matrix if and only if

$$a_1\,b_1 - a_2\,b_2 + a_3\,b_3 = 0.$$

- Except for the trivial structure the remaining parametric Poisson structures are all of rank 2 and two Casimirs are given by the quadratic functions

$$F_1(x) := a_1\,x_2^2 - a_2\,x_3^2 + a_3\,x_4^2, \qquad F_2(x) := a_1\,x_1^2 - b_3\,x_3^2 + b_2\,x_4^2.$$

Example 2.7 (*Momentum map on* $\mathfrak{e}(3)$)

The momentum map generalizes the classical notions of linear and angular momentum. To motivate the term "momentum map", consider a physical system of N particles moving in $\mathbb{R}^3$.

- The corresponding phase space is the manifold $\mathcal{M} = (T^*\mathbb{R}^3)^N \simeq \mathbb{R}^{6N}$ with canonical coordinates $(q, p) = ((q^{(1)}, p^{(1)}), \dots, (q^{(N)}, p^{(N)}))$, where

$$\left(q^{(i)}, p^{(i)}\right) = \left(q_1^{(i)}, q_2^{(i)}, q_3^{(i)}, p_1^{(i)}, p_2^{(i)}, p_3^{(i)}\right) \in \mathbb{R}^6, \qquad i = 1, \dots, N.$$

- The Hamiltonian of the system is a function $H \in \mathscr{F}(\mathbb{R}^{6N})$ which assigns to each point in the phase space the total energy of the system in that state. It determines the dynamics of the system as follows. A physical observable is a smooth function $F \in \mathscr{F}(\mathbb{R}^{6N})$. Its time-evolution is described by $\dot{F} = \{F, H\}$.
- Let $\mathbf{E}(3)$ be the Euclidean group on $\mathbb{R}^3$. Its action on $\mathbb{R}^3$ is given by

$$x \mapsto h\,x + y, \qquad h \in \mathbf{SO}(3),\, y \in \mathbb{R}^3.$$

- Consider the induced action of $\mathbf{E}(3)$ on $\mathcal{M}$. This action is Hamiltonian. The physically relevant actions are those that preserve the Hamiltonian H. In this example, if the Hamiltonian is preserved by $\mathbf{E}(3)$ then the dynamics does not depend on the position or the orientation of the N particle system as a whole. In other words, no external forces act on the system.
- The momentum map can be explicitly written as

$$\mu(q, p) = \sum_{i=1}^{N} \left(p^{(i)}, q^{(i)} \times p^{(i)}\right) \in \mathbb{R}^3 \times \mathfrak{so}(3)^*,$$

where the first component is the total linear momentum of the system and the second component is the total angular momentum of the system. One can easily see that the total derivative w.r.t. time of the momentum map vanishes, so that, as expected, the total linear and angular momentum of the system are conserved quantities.

Example 2.8 (*The two-body problem*)

As an application of "Noether Theorem" we consider the classical two-body problem, which is a conservative system of two point masses m_1 and m_2 in $\mathbb{R}^3$, where the interaction potential $U(q_1, q_2)$ depends only on the vector that joins the particles, $U(q_1, q_2) = U(Q)$, where $Q := q_1 - q_2 \in \mathbb{R}^3$.

- The canonical Hamiltonian is given by

$$H(q_1, q_2, p_1, p_2) := \frac{1}{2}\left(\frac{\|p_1\|^2}{m_1} + \frac{\|p_2\|^2}{m_2}\right) + U(Q),$$

which is a function on $T^*\mathbb{R}^6 \simeq \mathbb{R}^{12}$, equipped with the canonical symplectic structure.
- Clearly, H is invariant w.r.t. the natural translation action of $\mathbb{R}^3$, given by

$$(q_1, p_1, q_2, p_2) \mapsto (q_1 + a, p_1, q_2 + a, p_2), \qquad a \in \mathbf{G} = \mathbb{R}^3.$$

- The fundamental vector fields of this action are given by

$$\dot{q}_1 = \dot{q}_2 = \mathsf{X}, \qquad \dot{p}_1 = \dot{p}_2 = 0, \qquad \mathsf{X} \in \mathfrak{g} = \mathbb{R}^3.$$

- It follows that the linear map

$$F : \mathfrak{g} \to \mathscr{F}(\mathbb{R}^3) \,:\, \mathsf{X} \mapsto \langle \mathsf{X}, p_1 + p_2 \rangle,$$

is a co-momentum map for the action. Here $\langle \cdot, \cdot \rangle$ denotes the standard inner product on $\mathbb{R}^3$. Indeed,

$$\mathsf{v}_{\langle \mathsf{X}, p_1 + p_2 \rangle}[q_i] = \mathsf{X}, \qquad \mathsf{v}_{\langle \mathsf{X}, p_1 + p_2 \rangle}[p_i] = 0, \qquad i = 1, 2.$$

By "Noether Theorem" the three components of the total linear momentum $P := p_1 + p_2$ are integrals of motion.

- We have that H separates into the sum of two terms, $H = H_{\text{cm}} + H'$, where

$$H_{\text{cm}} := \frac{\|P\|^2}{2(m_1 + m_2)}$$

and

$$H' := \frac{m}{2}\left\|\frac{p_1}{m_1} - \frac{p_2}{m_2}\right\| + U(Q),$$

where m is the reduced mass defined by

$$\frac{1}{m} := \frac{1}{m_1} + \frac{1}{m_2}.$$

- As P is constant the center of mass of the two particles moves in a linear way since it is governed by H_{cm}. Let us therefore ignore this part of the total energy and focus on H'. Assume that $U(Q)$ depends only on the distance of the particles, i.e., $U(Q) = U(\rho)$, where $\rho := \|Q\|$.
- As a consequence our Hamiltonian system admits a second symmetry, namely a spherical symmetry, which is associated with the action of the Lie group $\mathbf{SO}(3)$ on $\mathbb{R}^3$. We can apply again "Noether Theorem" finding that all three components of its angular momentum are preserved, and hence the reduced particle evolves in a plane $\mathbb{R}^2$, with coordinates (ρ, θ). The conserved momentum in that plane, $\ell := m\,\rho^2\,\theta$, may be used to eliminate θ, leaving us with a one-degree of freedom Hamiltonian whose potential energy is

$$U_{\text{eff}}(\rho) = U(\rho) + \frac{\ell^2}{2\,m\,\rho}.$$

2.2 Symplectic manifolds

► We now introduce a special regular Poisson manifold which plays an important role in the theory of Hamiltonian dynamics when the phase space has maximal rank for any point. It turns out that such a manifold is locally diffeomorphic to the canonical Hamiltonian phase space $\mathbb{R}^{2n}$.

Definition 2.5

Let $\mathcal{M}$ be a real r-dimensional smooth manifold.

1. *If $\mathcal{M}$ is equipped with a closed non-degenerate 2-form $\omega \in \Omega^2(\mathcal{M})$ then the pair $(\mathcal{M}, \omega)$ is called* **symplectic manifold***.*
2. *Let $(\mathcal{M}, \omega)$ be a symplectic manifold and let $H \in \mathscr{F}(\mathcal{M})$. The vector field v_H defined as*

$$\mathsf{v}_H \lrcorner\, \omega = \omega(\mathsf{v}_H, \cdot) := \mathrm{d}H(\cdot), \tag{2.22}$$

is called the **Hamiltonian vector field** *associated with the Hamiltonian H. Its local flow $\Phi_t : \mathcal{I}_\varepsilon \times \mathcal{M} \to \mathcal{M}$, $\mathcal{I}_\varepsilon := \{t \in \mathbb{R} : |t| < \varepsilon\}$, is a* **Hamiltonian flow** *on $\mathcal{M}$.*

► From the above definition we can derive the general features of Hamiltonian dynamics on symplectic manifolds. We will see that a symplectic manifold is, in a natural way, an even-dimensional regular Poisson manifold with maximal rank on $\mathcal{M}$.

- Let us consider a coordinate chart $\mathcal{U}$ of $\mathcal{M}$ where a point $p \in \mathcal{M}$ can be expressed by local coordinates $x := (x_1, \ldots, x_r)$. Then ω can be locally expressed as
$$\omega = \sum_{1 \leqslant i < j \leqslant r} f_{ij}(x)\, \mathrm{d}x_i \wedge \mathrm{d}x_j,$$
for some smooth functions $f_{ij} \in \mathscr{F}(\mathcal{U})$.
  (a) The closure condition on ω, i.e., $\mathrm{d}\omega = 0$, means that locally there holds
  $$\frac{\partial f_{ij}}{\partial x_k} + \frac{\partial f_{jk}}{\partial x_i} + \frac{\partial f_{ki}}{\partial x_j} = 0 \qquad \forall\, i, j, k = 1, \ldots, r,$$
  and it assures that the induced Poisson bracket satisfies the Jacobi identity and the Hamiltonian flows define Poisson maps.
  (b) The non-degeneracy condition on ω means that locally there holds
  $$\mathsf{rank}\big(f_{ij}(x)\big)_{1 \leqslant i < j \leqslant r} = r.$$
  Such a condition, together with the skew-symmetry property of ω, implies that any symplectic manifold has even dimension, say $r = 2n$. Furthermore, the differential $2n$-form
  $$\omega^{2n} := \underbrace{\omega \wedge \cdots \wedge \omega}_{n \text{ times}} \tag{2.23}$$
  defines a volume form. In particular, every symplectic manifold is orientable.

- Formula (2.22) defining a Hamiltonian vector field on $(\mathcal{M}, \omega)$, implies that the 1-form $\mathsf{v}_H \lrcorner\, \omega$ is closed (even exact) for any vector field v_H, i.e.,
$$\mathsf{d}(\mathsf{v}_H \lrcorner\, \omega) = 0. \tag{2.24}$$

- The closure condition on ω implies a natural definition of a Poisson bracket on $\mathcal{M}$.
  (a) Let $F, H \in \mathscr{F}(\mathcal{M})$ and define the corresponding Hamiltonian vector fields $\mathsf{v}_F, \mathsf{v}_H$ by using (2.22). Then we define a skew-symmetric biderivation on $\mathscr{F}(\mathcal{M})$ by
  $$\{\, F, H \,\} := \omega(\mathsf{v}_F, \mathsf{v}_H) = \mathsf{d}F(\mathsf{v}_H) = \mathsf{v}_H[F]. \tag{2.25}$$
  The above formula shows that the definition of a Hamiltonian vector field on $\mathcal{M}$, see (2.22), is consistent with the definition of a Hamiltonian vector field on a Poisson manifold, see (2.1).

(b) The non-degeneracy and closure conditions of ω imply that (2.25) is a Poisson structure on $\mathcal{M}$.

Proof. Let $\mathsf{v}_F, \mathsf{v}_H$ be two Hamiltonian vector fields. Using $\mathrm{d}\omega = 0$ and formulas (1.29), (2.22), (2.24) and (2.25) we get

$$\begin{aligned} [\mathsf{v}_F, \mathsf{v}_H] \lrcorner\, \omega &= \mathrm{d}(\mathsf{v}_F \lrcorner\, \mathsf{v}_H \lrcorner\, \omega) \\ &\quad + \mathsf{v}_F \lrcorner\, \mathrm{d}(\mathsf{v}_H \lrcorner\, \omega) - \mathsf{v}_H \lrcorner\, \mathrm{d}(\mathsf{v}_F \lrcorner\, \omega) - \mathsf{v}_H \lrcorner\, \mathsf{v}_F \lrcorner\, \mathrm{d}\omega \\ &= \mathrm{d}(\mathsf{v}_F \lrcorner\, \mathsf{v}_H \lrcorner\, \omega) = -\mathrm{d}(\omega(\mathsf{v}_F, \mathsf{v}_H)) \\ &= -\mathrm{d}\{F, H\} = -\mathsf{v}_{\{F,H\}} \lrcorner\, \omega. \end{aligned}$$

Since ω is non-degenerate this shows that

$$[\mathsf{v}_F, \mathsf{v}_H] = -\mathsf{v}_{\{F,H\}}. \tag{2.26}$$

Applying (2.26) to an arbitrary $G \in \mathscr{F}(\mathcal{M})$ we find

$$\mathsf{v}_F\,[\mathsf{v}_H[G]] - \mathsf{v}_H\,[\mathsf{v}_F[G]] + \mathsf{v}_{\{F,H\}}[G] = 0,$$

which is precisely the Jacobi identity, as follows from (2.25).

(c) A more explicit expression for the non-degenerate Poisson bracket induced by ω can be easily derived. Picking local coordinates $(x_1, \ldots, x_r)$, $r = 2n$, on a chart $\mathcal{U}$ we can introduce a skew-symmetric $r \times r$ matrix $\Omega := (\Omega_{ij})_{1 \leqslant i,j \leqslant r}$ by setting

$$\Omega_{ij} := \omega\left(\frac{\partial}{\partial x_i}, \frac{\partial}{\partial x_j}\right). \tag{2.27}$$

Let $F, H \in \mathscr{F}(\mathcal{U})$. Denote by $[\mathsf{v}_F]$ the column matrix whose elements are the coefficients of v_F w.r.t. the basis $\{\partial/\partial x_1, \ldots, \partial/\partial x_r\}$ and by $[\mathrm{d}F]$ the column matrix whose elements are the coefficients of $\mathrm{d}F$ w.r.t. the basis $\{\mathrm{d}x_1, \ldots, \mathrm{d}x_n\}$. Then (2.22) says that $\Omega[\mathsf{v}_F] = -[\mathrm{d}F]$ and $\omega(\mathsf{v}_F, \mathsf{v}_H) = [\mathsf{v}_F]^\top \Omega [\mathsf{v}_H]$, so that

$$\begin{aligned} \{F, G\} &= [\mathsf{v}_F]^\top \Omega\, [\mathsf{v}_H] = \left(\Omega^{-1}[\mathrm{d}F]\right)^\top \Omega \left(\Omega^{-1}[\mathrm{d}H]\right) \\ &= -[\mathrm{d}F]^\top \Omega^{-1} [\mathrm{d}H], \end{aligned}$$

which compared with (2.9) gives the Poisson matrix $\mathcal{X}$ corresponding to (2.25):

$$\mathcal{X} = -\Omega^{-1}. \tag{2.28}$$

- Any flow of a Hamiltonian vector field v_H on $(\mathcal{M}, \omega)$ leaves ω invariant (as well as the induced Poisson bracket, in view of (2.25), and the volume form (2.23)). Equivalently, if Φ_t is the Hamiltonian flow of v_H then $\mathfrak{L}_{\mathsf{v}_H}\omega = 0$.

Proof. By using Cartan formula (1.27) and $d\omega = 0$, we have:

$$\mathfrak{L}_{v_H}\omega = d(v_H \lrcorner \omega) + v_H \lrcorner d\omega = d(v_H \lrcorner \omega) = 0,$$

in view of (2.24).

- An equivalent way to claim that any Hamiltonian vector field v_H on $(\mathcal{M}, \omega)$ leaves ω invariant is to say that any Hamiltonian flow Φ_t is a *symplectic morphism*, which is the natural generalization to manifolds of symplectic transformations in canonical Hamiltonian mechanics.

► As anticipated, any symplectic manifold is locally diffeomorphic to the canonical Hamiltonian phase space $\mathbb{R}^{2n}$ presented in Example 2.9. This is the content of the following Theorem, for which we omit the proof.

Theorem 2.2 (*Darboux*)

Let $(\mathcal{M}, \omega)$ be a $2n$-dimensional symplectic manifold. Then for every $p \in \mathcal{M}$ there exists a coordinate chart $\mathcal{U}$ around p and a diffeomorphism $\varphi : \mathcal{U} \to \mathcal{V}$, where $\mathcal{V}$ is an open subset of $\mathbb{R}^{2n}$ parametrized by $(q_1, \ldots, q_n, p_1, \ldots, p_n)$ such that

$$\varphi^\star \left(\sum_{j=1}^{n} dq_j \wedge dp_j \right) = \omega$$

on $\mathcal{U}$.

No Proof.

2.2.1 *Examples*

► Let us present some examples of symplectic manifolds.

Example 2.9 (*The canonical Hamiltonian phase space* $\mathbb{R}^{2n}$)

We already know that the canonical Hamiltonian phase space

$$\left(\mathbb{R}^{2n}, \omega_{\mathrm{can}}\right), \qquad \omega_{\mathrm{can}} := \sum_{j=1}^{n} dq_j \wedge dp_j$$

is a symplectic vector space (of dimension $r = 2n$).

- If $H \in \mathscr{F}(\mathbb{R}^{2n})$ then, by using (2.22), we can construct the corresponding canonical Hamiltonian vector field. The differential of H is

$$dH = \sum_{i=1}^{n} \left(\frac{\partial H}{\partial q_i} dq_i + \frac{\partial H}{\partial p_i} dp_i \right),$$

and a vector field on $\mathcal{M}$ can be written as

$$v = \sum_{i=1}^{n} \left(f_i(q,p) \frac{\partial}{\partial q_i} + g_i(q,p) \frac{\partial}{\partial p_i} \right).$$

Then, v is the Hamiltonian vector field corresponding to H is and only if

$$\begin{aligned} \mathrm{d}H &= \sum_{i=1}^{n} \left(\frac{\partial H}{\partial q_i} \mathrm{d}q_i + \frac{\partial H}{\partial p_i} \mathrm{d}p_i \right) \\ &= \left(\sum_{i=1}^{n} \left(f_i(q,p) \frac{\partial}{\partial q_i} + g_i(q,p) \frac{\partial}{\partial p_i} \right) \right) \lrcorner \left(\sum_{k=1}^{n} \mathrm{d}q_k \wedge \mathrm{d}p_k \right), \end{aligned}$$

which gives the canonical Hamiltonian vector field

$$f_i(q,p) = \frac{\partial H}{\partial p_i}, \qquad g_i(q,p) = -\frac{\partial H}{\partial q_i}, \qquad i = 1, \dots, n,$$

as expected. In particular, we find

$$\mathsf{v}_{q_i} = -\frac{\partial}{\partial p_i}, \qquad \mathsf{v}_{p_i} = \frac{\partial}{\partial q_i}, \qquad i = 1, \dots, n.$$

- The *canonical Poisson structure* on $\mathbb{R}^{2n}$ is defined in terms of (2.27):

$$\{ q_i, p_j \} := \omega_{\mathrm{can}} \left(\mathsf{v}_{q_i}, \mathsf{v}_{p_j} \right) = \left(\sum_{k=1}^{n} \mathrm{d}q_k \wedge \mathrm{d}p_k \right) \left(-\frac{\partial}{\partial p_i}, \frac{\partial}{\partial q_j} \right) = \delta_{ij},$$

for $i,j = 1, \dots, n$. Similarly, $\{q_i, q_j\} = \{p_i, p_j\} = 0$ for any $i,j = 1, \dots, n$. Therefore, see (2.28), the (non-degenerate) Poisson matrix corresponding to ω_{can} is

$$\mathcal{X} = -\mathcal{X}_{\mathrm{can}}^{-1} = \mathcal{X}_{\mathrm{can}},$$

where

$$\mathcal{X}_{\mathrm{can}} := \begin{pmatrix} 0_n & 1_n \\ -1_n & 0_n \end{pmatrix}.$$

Example 2.10 (*The cotangent bundle of a smooth manifold*)

Let $\mathcal{M}$ be a n-dimensional smooth manifold and let $T^*\mathcal{M}$ be its ($2n$-dimensional) cotangent bundle. It can be proved that $T^*\mathcal{M}$ is a symplectic manifold. Note that this implies that $T^*\mathcal{M}$ is an orientable manifold even if $\mathcal{M}$ is not.

- Suppose that $(x_1, \dots, x_n)$ are local coordinates on a neighborhood $\mathcal{U}$. They give half of a system of coordinates on $T^*\mathcal{U} \subset T^*\mathcal{M}$ by letting $q_i := x_i \circ \pi$ where $\pi : T^*\mathcal{M} \to \mathcal{M}$ is the natural projection from the cotangent bundle to $\mathcal{M}$.
- For the other half let p_i denote the function on $T^*\mathcal{U}$ defined by

$$p_i(y) := \left\langle y, \left. \frac{\partial}{\partial x_i} \right|_p \right\rangle, \qquad i = 1, \dots, n,$$

where $y \in T^*\mathcal{U}$ is in the fiber over $p \in \mathcal{M}$, i.e., $\pi(y) = p$. Therefore our coordinates on $T^*\mathcal{U}$ are $(q_1, \dots, q_n, p_1, \dots, p_n)$.

- On $T^*\mathcal{M}$ there is a natural differential 1-form, called *Liouville form*, defined by

$$\vartheta := \sum_{i=1}^{n} p_i \, \mathrm{d}q_i.$$

More intrinsically, $\vartheta \in \Omega(T^*\mathcal{M})$ is defined by

$$\langle \vartheta, \mathsf{v} \rangle (y) := \langle y, T_y\pi(\mathsf{v}(y)) \rangle,$$

where $y \in T^*\mathcal{M}$ and $\mathsf{v} \in \mathfrak{X}(T^*\mathcal{M})$. The canonical pairing in this formula is between $T_p^*\mathcal{M}$ and $T_p\mathcal{M}$: indeed, $T_y\pi : T_y(T^*\mathcal{M}) \to T_p\mathcal{M}$.

- The canonical symplectic structure on $T^*\mathcal{M}$ is by definition the closed differential 2-form

$$\omega_{\text{can}} := -\mathsf{d}\vartheta = \sum_{i=1}^{n} \mathsf{d}q_i \wedge \mathsf{d}p_i.$$

- A standard way to present classical mechanics on symplectic manifolds is to start with the assumption that the configuration space of a system is a n-dimensional smooth manifold $\mathcal{M}$. Then the natural *Lagrangian phase space* is the $2n$-dimensional tangent bundle $T\mathcal{M}$, while the natural *Hamiltonian phase space* is the $2n$-dimensional cotangent bundle $T^*\mathcal{M}$. In this setting one can formulate both Lagrangian and Hamiltonian mechanics in a very natural way.
- In particular, a *Hamiltonian system* is defined in terms of a smooth Hamiltonian

$$H : T^*\mathcal{M} \to \mathbb{R}.$$

In any coordinate chart of $T^*\mathcal{M}$ with coordinates $(q_1, \ldots, q_n, p_1, \ldots, p_n)$ the equations of motion are given by the *canonical Hamilton equations*

$$\begin{cases} \dot{q}_i = \dfrac{\partial H}{\partial p_i}, \\[2ex] \dot{p}_i = -\dfrac{\partial H}{\partial q_i}, \end{cases}$$

with $i = 1, \ldots, n$.

Example 2.11 (*The sphere* $\mathbb{S}^2$)

The 2-sphere $\mathbb{S}^2$, regarded as the set of unit vectors in $\mathbb{R}^3$, has tangent vectors at $p \in \mathbb{S}^2$ identified with vectors orthogonal to the vector connecting the origin with p (see Example 1.6).

- The standard symplectic form on $\mathbb{S}^2$ is induced by the standard scalar and vector products:

$$(\omega(\mathsf{v}, \mathsf{w}))(p) := \langle\, p, \mathsf{v} \times \mathsf{w}\,\rangle\,, \tag{2.29}$$

where $p \in \mathbb{S}^2$ and $\mathsf{v}, \mathsf{w} \in T_p\mathbb{S}^2$. This is the standard *area form* on $\mathbb{S}^2$ with total area $4\,\pi$.
- In terms of cylindrical coordinates $(\theta, h) \in [0, 2\,\pi) \times [-1, 1]$ away from the poles, one finds $\omega = \mathsf{d}\theta \wedge \mathsf{d}h$. The vector fiels $\partial/\partial\theta$ is a Hamiltonian vector field with Hamiltonian $H = h$:

$$\frac{\partial}{\partial\theta} \lrcorner\ (\mathsf{d}\theta \wedge \mathsf{d}h) = \mathsf{d}h.$$

The motion generated by this vector field corresponds to a rotation about the vertical axis, which of course preserves both area and height.

It can be proved that any oriented two-dimensional smooth manifold with an area form is a symplectic manifold.

2.3 Foliation of a Poisson manifold

► Before considering the problem of the foliation of a Poisson manifold equipped with a degenerate Poisson structure we consider the simpler case when the Poisson manifold has maximal rank everywhere, which implies that its dimension is necessarily even and that $\mathsf{Ham}_p(\mathcal{M}) = T_p\mathcal{M}$ for all points $p \in \mathcal{M}$. In view of what we know about symplectic manifolds the next claim is not surprising.

Theorem 2.3

Let $(\mathcal{M}, \{\cdot,\cdot\})$ be a regular Poisson manifold with maximal rank on $\mathcal{M}$. Then $\mathcal{M}$ is a symplectic manifold.

Proof. We proceed by steps.

- Due to the maximal rank assumption on $\mathcal{M}$ the map $\widetilde{\mathsf{P}} : \Omega(\mathcal{M}) \to \mathfrak{X}(\mathcal{M})$ induced by the Poisson 2-vector field P, $\mathsf{P}(\mathrm{d}F, \mathrm{d}H) := \{F, H\}$, $F, H \in \mathscr{F}(\mathcal{M})$, is everywhere invertible.

- We define ω to be the 2-form which corresponds to its inverse $\widetilde{\mathsf{P}}^{-1}$. Since $\widetilde{\mathsf{P}}(\mathrm{d}H) = \mathsf{v}_H$, for any $H \in \mathscr{F}(\mathcal{M})$, see (2.2), we define the 2-form

$$\omega\left(\mathsf{v}_F, \mathsf{v}_H\right) := \left(\widetilde{\mathsf{P}}^{-1}(\mathsf{v}_F)\right)(\mathsf{v}_H) = \mathrm{d}F(\mathsf{v}_H) = \{F, H\}, \tag{2.30}$$

  for any $F, H \in \mathscr{F}(\mathcal{M})$.

- Obviously, ω is non-degenerate due to the non-degeneracy of the Poisson structure on $\mathcal{M}$. The closure condition is a consequence of the Jacobi identity. Indeed, applying formula (1.18) to three Hamiltonian vector fields $\mathsf{v}_F, \mathsf{v}_G, \mathsf{v}_H$ we get

$$\begin{aligned} \mathrm{d}\omega(\mathsf{v}_F, \mathsf{v}_G, \mathsf{v}_H) &= \mathsf{v}_F[\omega(\mathsf{v}_G, \mathsf{v}_H)] + \omega(\mathsf{v}_F, [\mathsf{v}_G, \mathsf{v}_H]) + \circlearrowleft (\mathsf{v}_F, \mathsf{v}_G, \mathsf{v}_H) \\ &= \mathsf{v}_F[\{G, H\}] - \omega(\mathsf{v}_F, \mathsf{v}_{\{G,H\}}) + \circlearrowleft (F, G, H) \\ &= \{\{G, H\}, F\} - \{F, \{G, H\}\} + \circlearrowleft (F, G, H) \\ &= 2(\{\{G, H\}, F\} + \circlearrowleft (F, G, H)) \\ &= 0, \end{aligned}$$

  where we used (2.4), (2.30) and the Jacobi identity. Since $T_p\mathcal{M} = \mathsf{Ham}_p(\mathcal{M})$ for any $p \in \mathcal{M}$ (by the maximal rank assumption) this shows that ω is closed and hence that ω is a symplectic form.

The Theorem is proved. ■

► Note that combining Theorem 2.2 with Theorem 2.3 we get as a corollary that if $(\mathcal{M}, \{\cdot,\cdot\})$ is a regular Poisson manifold with maximal rank on $\mathcal{M}$ then $\mathcal{M}$ is locally diffeomorphic to the canonical Hamiltonian phase space $\mathbb{R}^{2n}$ with symplectic form given by

$$\omega_{\text{can}} := \sum_{j=1}^{n} \mathrm{d}q_j \wedge \mathrm{d}p_j.$$

► We now describe the Poisson structure in the neighborhood of any point of a Poisson manifold and deduce from it a (global) decomposition, called *symplectic foliation* (or *symplectic stratification*) of the Poisson manifold into *symplectic leaves* (or *symplectic strata*) (of varying dimensions). The foliation of a Poisson manifold is described by the next Theorem, which is a generalization of Theorem 2.3 valid for regular Poisson manifolds. Let us mention that the claim that an arbitrary Poisson manifold admits a foliation whose strata are symplectic manifolds goes back to Lie.

Theorem 2.4 (*Weinstein*)

Let $(\mathcal{M}, \{\cdot,\cdot\})$ be a n-dimensional Poisson manifold and let $p \in \mathcal{M}$. Assume that

$$\mathsf{rank}_p\{\cdot,\cdot\} = 2r.$$

Then there exists a coordinate chart $\mathcal{U}$ of p with coordinates

$$(q_1, \ldots, q_r, p_1, \ldots, p_r, z_1, \ldots, z_s), \qquad s := n - 2r,$$

such that the Poisson 2-vector field on $\mathcal{U}$ takes the form

$$\mathsf{P} = \sum_{i=1}^{r} \frac{\partial}{\partial q_i} \wedge \frac{\partial}{\partial p_i} + \sum_{1 \leqslant i < j \leqslant s} f_{ij}(z_1, \ldots, z_s) \frac{\partial}{\partial z_i} \wedge \frac{\partial}{\partial z_j}, \tag{2.31}$$

where any f_{ij} is a smooth function which vanishes at p.

Proof. For $r = 0$ it is clear that for every Poisson manifold $(\mathcal{M}, \{\cdot,\cdot\})$ and for every point p such that the rank of $\{\cdot,\cdot\}$ at p is zero, an arbitrary system of local coordinates $(z_1, \ldots, z_n)$ centered ad p works. Let $r = 1$. We show the existence of coordinates $(q_1, p_1, z_1, \ldots, z_{n-2})$ on a neighborhood $\mathcal{U}$ of p, for which (2.31) holds, where the functions f_{ij} are smooth and depend on $z_1, \ldots, z_{n-2}$ only. The proof then follows by induction on r.

- Fix $r = 1$. There exists a function p_1 such that $\mathsf{v}_{p_1}(p) \neq 0$. By the "Straightening Theorem" there exists a neighborhood $\mathcal{V}$ of p and a function q_1 on it, such that $\mathsf{v}_{p_1} = \partial/\partial q_1$ on $\mathcal{V}$. Notice that $\{q_1, p_1\} = \mathsf{v}_{p_1}[q_1] = \partial q_1/\partial q_1 = 1$.

- It follows that v_{p_1} and v_{q_1} define an integrable (in the sense of Frobenius) distribution of rank 2 on a neighborhood $\mathcal{V}$ of p: the vector fields v_{p_1} and v_{q_1} are independent on a neighborhood of p because they are independent at p, and the vector space spanned by these vector fields forms a Lie subalgebra of $\mathfrak{X}(\mathcal{V})$ since

$$[\mathsf{v}_{p_1}, \mathsf{v}_{q_1}] = \mathsf{v}_{\{q_1, p_1\}} = \mathsf{v}_1 = 0.$$

- By the "Frobenius Theorem" the vector fields v_{p_1} and v_{q_1} define locally a foliation with 2-dimensional leaves. On a neighborhood $\mathcal{U}$ of p we can then find functions $z_1, \ldots, z_{n-2}$ such that $\mathrm{d}z_1, \ldots, \mathrm{d}z_{n-2}$ are independent on $\mathcal{U}$ and such that $\mathsf{v}_{p_1}[z_i] = \mathsf{v}_{q_1}[z_i] = 0$, $i = 1, \ldots, n-2$.

- It follows that $q_1, p_1, z_1, \dots, z_{n-2}$ are coordinates on $\mathcal{U}$ and that the Poisson 2-vector field on $\mathcal{U}$ takes the form

$$\mathsf{P} = \frac{\partial}{\partial q_1} \wedge \frac{\partial}{\partial p_1} + \sum_{1 \leqslant i,j \leqslant n-2} \{ z_i, z_j \} \frac{\partial}{\partial z_i} \wedge \frac{\partial}{\partial z_j}.$$

- In order to show that $\{ z_i, z_j \}$ is independent of q_1 and p_1 it suffices to show that $\{\{ z_i, z_j \}, p_1\} = 0$ and $\{\{ z_i, z_j \}, q_1\} = 0$, which are consequences of the Jacobi identity.

The Theorem is proved. ■

► Remarks:

- The rank of $\{ \cdot, \cdot \}$ at p is $2r$ but is not necessarily constant on a neighborhood $\mathcal{U}$ of p. When the rank is constant and equal to $2r$ on $\mathcal{U}$, then $\mathcal{U}$ can be chosen in such a way that the functions f_{ij} vanish on $\mathcal{U}$. This leads to the situation described by Theorem 2.3 and the Poisson 2-vector field is the canonical one:

$$\mathsf{P}_{\text{can}} = \sum_{i=1}^{r} \frac{\partial}{\partial q_i} \wedge \frac{\partial}{\partial p_i}.$$

- If the rank of $\{ \cdot, \cdot \}$ at p is constant and equal to $2r$ on a neighborhood $\mathcal{U}$ of p, then the Hamiltonian vector fields on $\mathcal{U}$ define a distribution of rank $2r$. Such a distribution is integrable (in the sense of Frobenius) because $[\, \mathsf{v}_F, \mathsf{v}_G \,] = -\mathsf{v}_{\{F,G\}}$ for all $F, G \in \mathscr{F}(\mathcal{U})$. Thus, we have $n - 2r$ Casimirs and a (regular) foliation of $\mathcal{U}$, where each leaf has dimension $2r$. The leaves inherit a Poisson structure from $\{ \cdot, \cdot \}$. Since the rank of this Poisson structure is $2r$ each leaf carries a symplectic form and $\mathcal{U}$ admits a decomposition into *symplectic leaves*.

- Remarkably, the decomposition into symplectic leaves exists in the neighborhood of any point p of a Poisson manifold. In other words, the rank needs not be constant in a neighborhood of p, but the dimension of the leaves will not be constant in general. Indeed, one can prove that if $(\mathcal{M}, \{ \cdot, \cdot \})$ is a Poisson manifold then the (singular) distribution on $\mathcal{M}$ defined by the Hamiltonian vector fields is integrable in the sense that every $p \in \mathcal{M}$ has a coordinate neighborhood $\mathcal{U}$ which is, in a unique way, a disjoint union of symplectic manifolds $\mathcal{M}_\alpha$ which are Poisson submanifolds of $\mathcal{U}$. The resulting foliation is called the *symplectic foliation* of $\mathcal{M}$ and each of its leaves (of varying dimensions) is called *symplectic leaf*.

- Let $\mathcal{M}_\alpha \subset \mathcal{M}$ be a symplectic leaf of $\mathcal{M}$.

(a) Theorem 2.3 guarantees that we can construct a symplectic 2-form ω on $\mathcal{M}_\alpha$ by setting

$$\omega(\mathsf{v}_F, \mathsf{v}_H) := \{ F, H \},$$

for any two Hamiltonian vector fields $\mathsf{v}_F, \mathsf{v}_H$ on $\mathcal{M}_\alpha$. Note that two points belong to the same symplectic leaf if they can be joined by a path whose tangent vector at any point is a Hamiltonian vector field.

(b) If $F \in \mathscr{F}(\mathcal{M})$ is a Casimir function then F is constant on $\mathcal{M}_\alpha$.

Proof. If F were not locally constant on $\mathcal{M}_\alpha$, then there would exist a point $p \in \mathcal{M}_\alpha$ such that $(\mathrm{d}F(\mathsf{v}))(p) \neq 0$ for some $\mathsf{v} \in T_p\mathcal{M}_\alpha$. But $T_p\mathcal{M}_\alpha$ is spanned by v_H for $H \in \mathscr{F}(\mathcal{M})$ and hence $(\mathsf{v}_H[F])(p) = \{F, H\}(p) = 0$ which implies $(\mathrm{d}F(\mathsf{v}))(p) = 0$ which is a contradiction. Thus F is locally constant on a neighborhood of p and hence constant by connectedness of the leaf $\mathcal{M}_\alpha$.

In good cases most or all of the symplectic leaves are level sets of the Casimir functions, but this is not true in general. One can prove that locally, the regular leaves (i.e., the leaves which pass through a point where the rank is maximal, hence locally constant) are given as the level sets of the (local) Casimir functions.

Example 2.12 (*The Euler top and the symplectic foliation of its phase space*)

The three-dimensional Euler top described in Example 2.4 is a simple but non-trivial example of a mechanical system on a Poisson manifold whose phase space admits a symplectic foliation.

- We know that the system is governed by the ODEs

$$\begin{cases} \dot{x}_1 = (a_2 - a_3)\, x_2\, x_3, \\ \dot{x}_2 = (a_3 - a_1)\, x_3\, x_1, \\ \dot{x}_3 = (a_1 - a_2)\, x_1\, x_2, \end{cases} \tag{2.32}$$

where $a_1, a_2, a_3 \in \mathbb{R}$.

- This system is a Hamiltonian system w.r.t. the Poisson structure

$$\{ x_i, x_j \} = \varepsilon_{ijk}\, x_k, \tag{2.33}$$

and its admits two functionally independent and Poisson involutive integrals of motion: the Hamiltonian

$$H(x) := \frac{1}{2}\left(a_1\, x_1^2 + a_2\, x_2^2 + a_3\, x_3^2\right),$$

and the Casimir

$$B(x) := \frac{1}{2}\left(x_1^2 + x_2^2 + x_3^2\right).$$

- The level sets

$$\mathcal{S}_b := \left\{x \in \mathbb{R}^3 \,:\, B(x) = b^2\right\},$$

with $b \geqslant 0$ are the symplectic leaves of the Poisson manifold. These are two-dimensional spheres which can be equipped with a symplectic structure. On each symplectic leaf $\mathcal{S}_b$ the Hamiltonian dynamics is locally canonical.

- Excluding the origin (i.e., $b = 0$), which provides a singular leaf, the symplectic 2-form on $\mathcal{S}_b$ is given by (see (2.29))

$$(\omega(\mathsf{v}, \mathsf{w}))(p) := \frac{1}{b^2}\, \langle\, p, \mathsf{v} \times \mathsf{w}\,\rangle,$$

where $p \in S^2$ and $\mathsf{v}, \mathsf{w} \in T_p S_b$. Note that the Poisson bracket (2.33) induces on each S_b a well-defined symplectic Hamiltonian vector field.

- Since both H and B are integrals of motion, the motion takes place along the intersections of the level surfaces of the energy H (ellipsoids embedded in $\mathbb{R}^3$) and the angular momentum B (spheres embedded in $\mathbb{R}^3$). The centers of the energy ellipsoids and the angular momentum spheres coincide. The two sets of level surfaces in $\mathbb{R}^3$ develop collinear gradients (for example, tangencies) at pairs of points that are diametrically opposite on an angular momentum sphere. At these points, collinearity of the gradients of H and B implies stationary rotations, that is, fixed points.

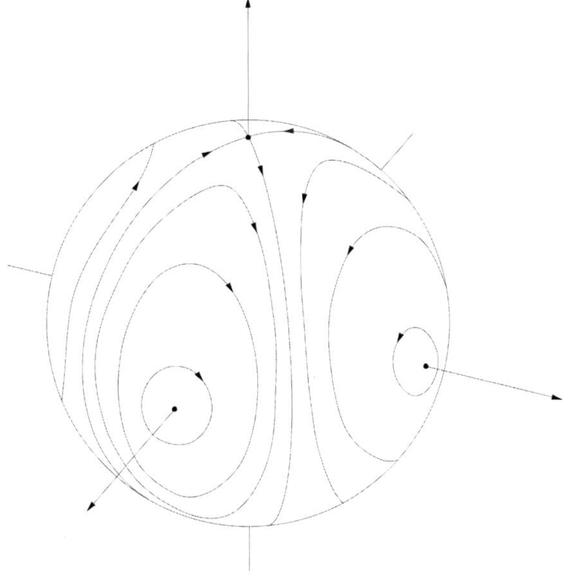

Fig. 4.5. Representation of Euler top flow on the angular momentum sphere.

Example 2.13 (*The open book foliation*)

Consider the linear Poisson structure on $\mathbb{R}^3$ corresponding to the Poisson matrix

$$\mathcal{X} := \begin{pmatrix} 0 & 0 & x_1 \\ 0 & 0 & x_2 \\ -x_1 & -x_2 & 0 \end{pmatrix}.$$

- On $\mathbb{R}^3 \setminus \{x_1 = 0\}$ the function x_2/x_1 is a Casimir, while on $\mathbb{R}^3 \setminus \{x_2 = 0\}$ the function x_1/x_2 is a Casimir.
- The symplectic leaves of dimension 0 are the points on the x_3-axis, while the symplectic leaves of dimension 2 are the half-planes, obtained by removing the x_3-axis from all the planes which pass through the x_3-axis. This symplectic foliation is called *open book foliation*. Observe that on no neighborhood of an arbitrary point on the x_3-axis there is a non-constant Casimir.

2.4 Construction of Poisson manifolds

► We now briefly describe three basic methods to construct new Poisson manifolds from given ones:

1. Construction of Poisson submanifolds of given Poisson manifolds.
2. Construction of product Poisson manifolds.
3. Poisson reductions defined by means of Poisson actions of Lie groups.

► We already noticed that a submanifold of a Poisson manifold is, in general, not a Poisson submanifold. We will give precise conditions for this to happen in the claims which follow.

Theorem 2.5

Let $(\mathcal{M}, \{\cdot,\cdot\})$ be a Poisson manifold and let $\mathcal{M}'$ be a (possibly immersed) submanifold. There exists a Poisson structure $\{\cdot,\cdot\}'$ for which $\mathcal{M}'$ is an immersed Poisson submanifold of $\mathcal{M}$ if and only if the restriction of every Hamiltonian vector field on $\mathcal{M}$ to $\mathcal{M}'$ is tangent to $\mathcal{M}'$.

No Proof.

► Remarks:

- The symplectic leaves of a Poisson manifold are immersed submanifolds to which all Hamiltonian vector fields are tangent. Therefore, Theorem 2.5 yields another proof that these leaves carry a Poisson structure.
- Suppose that $F_1, \ldots, F_s \in \mathscr{F}(\mathcal{M})$ are Casimir functions and that $c := (c_1, \ldots, c_s)$ are constants such that

$$\mathcal{A}_c := \bigcap_{i=1}^{s} \{p \in \mathcal{M} \, : \, F_i(p) = c_i\}$$

  is a (non-empty) submanifold of $\mathcal{M}$. Since $\mathsf{v}_H[F_i] = \{F_i, H\} = 0$ for all $H \in \mathscr{F}(\mathcal{M})$ and $i = 1, \ldots, s$, all Hamiltonian vector fields v_H are tangent to $\mathcal{A}_c$ and $\mathcal{A}_c$ is a Poisson submanifold of $\mathcal{M}$.
- Let $p \in \mathcal{M}$ and suppose that $\mathcal{M}'$ is a Poisson submanifold, passing through p. Then $\mathcal{M}'$ contains, at least in a neighborhood of p, the symplectic leaf of $\mathcal{M}$ which passes through p, so $\mathcal{M}'$ is locally a union of symplectic leaves.

► A second construction of new Poisson manifolds is provided by the product of some given Poisson manifolds.

Theorem 2.6

Let $(\mathcal{M}_i, \{\cdot,\cdot\}_i)$, $i = 1,2$, be two Poisson manifolds. The product $\mathcal{M}_1 \times \mathcal{M}_2$ has a natural Poisson bracket such that the two projection maps $\pi_i : \mathcal{M}_1 \times \mathcal{M}_2 \to \mathcal{M}_i$ are Poisson morphisms.

Proof. In order for π_1 and π_2 to be Poisson morphisms it is necessary and sufficient to define

$$\{\pi_i^\star F_i, \pi_i^\star G_i\} := \pi_i^\star \{F_i, G_i\}_i \qquad \forall\, F_i, G_i \in \mathscr{F}(\mathcal{M}_i),$$

with $i = 1,2$. We define in addition

$$\{\pi_1^\star F_1, \pi_2^\star F_2\} := 0 \qquad \forall\, F_1 \in \mathscr{F}(\mathcal{M}_1),\ F_2 \in \mathscr{F}(\mathcal{M}_2).$$

These definitions extend uniquely to a skew-symmetric biderivation on $\mathscr{F}(\mathcal{M}_1 \times \mathcal{M}_2)$ which we denote by $\{\cdot,\cdot\}$. Notice that the Poisson matrix of $\{\cdot,\cdot\}$ w.r.t. the system of local coordinates coming from local coordinates on $\mathcal{M}_1$ and $\mathcal{M}_2$ has a block form, where each block is the pull-back under $\pi_i^\star$ of the Poisson matrix w.r.t. those local coordinates on $\mathcal{M}_1$ and $\mathcal{M}_2$. Therefore, the Jacobi identity is satisfied. ■

► Remarks:

- The fact that the matrix of $\{\cdot,\cdot\}$ has block form shows that the rank of $\{\cdot,\cdot\}$ at $(p_1, p_2) \in \mathcal{M}_1 \times \mathcal{M}_2$ is given by the sum of the ranks, that is $\mathsf{rank}_{p_1}\{\cdot,\cdot\}_1 + \mathsf{rank}_{p_2}\{\cdot,\cdot\}_2$. In particular,

$$\mathsf{rank}\{\cdot,\cdot\} = \mathsf{rank}\{\cdot,\cdot\}_1 + \mathsf{rank}\{\cdot,\cdot\}_2.$$

- The above construction can be generalized to the product of several Poisson manifolds.
- In view of Theorem 2.6 one can restate Theorem 2.4 by saying: *Every point p in a Poisson manifold $(\mathcal{M}, \{\cdot,\cdot\})$ has a coordinate neighborhood which is the product of a symplectic manifold of dimension $\mathsf{rank}_p\{\cdot,\cdot\}$ and a Poisson manifold which has rank 0 at the point that corresponds to p.*

► A third construction consists of the simplest form of a *Poisson reduction*. It is useful to give the following definitions.

- Let $\mathbf{G}$ be a Lie group with multiplication $* : \mathbf{G} \times \mathbf{G} \to \mathbf{G}$. If $\{\cdot,\cdot\}_\mathbf{G}$ is a Poisson structure on $\mathbf{G}$ such that $*$ is a Poisson morphism then $(\mathbf{G}, \{\cdot,\cdot\}_\mathbf{G})$ is called *Lie-Poisson group*.
- Let $\mathbf{G}$ be a Lie group acting on a smooth manifold $\mathcal{M}$. The action $\varphi : \mathbf{G} \times \mathcal{M} \to \mathcal{M}$ is a *Poisson action* if φ is a Poisson morphism, where one considers the product bracket on $\mathbf{G} \times \mathcal{M}$.

► We now give the following claim. Note that we are not demanding that $\mathbf{G}$ is a Lie-Poisson group.

Theorem 2.7

> *Let* **G** *be a Lie group, equipped with a Poisson structure* $\{\cdot,\cdot\}_{\mathbf{G}}$*, acting on a Poisson manifold* $(\mathcal{M},\{\cdot,\cdot\})$ *by means of a Poisson action. Then the algebra* $\mathscr{F}_{\mathbf{G}}(\mathcal{M})$ *of* **G***-invariant functions is a Poisson subalgebra of* $\mathscr{F}(\mathcal{M})$*, i.e., it is closed under* $\{\cdot,\cdot\}$.

Proof. Let us denote the action $\mathbf{G}\times\mathcal{M}\to\mathcal{M}$ by φ and the projection $\mathbf{G}\times\mathcal{M}\to\mathcal{M}$ by π. By definition, $F\in\mathscr{F}(\mathcal{M})$ is **G**-invariant if and only if $F\circ\varphi=F\circ\pi$. Thus, if $F,G\in\mathscr{F}_{\mathbf{G}}(\mathcal{M})$ and φ is Poisson, then

$$\varphi^{\star}\{F,G\}_{\mathcal{M}}=\{\varphi^{\star}F,\varphi^{\star}G\}_{\mathbf{G}\times\mathcal{M}}=\{\pi^{\star}F,\pi^{\star}G\}_{\mathbf{G}\times\mathcal{M}}=\pi^{\star}\{F,G\}_{\mathcal{M}},$$

which means that the bracket of any two **G**-invariant functions is **G**-invariant. Therefore the subalgebra $\mathscr{F}_{\mathbf{G}}(\mathcal{M})$ of $\mathscr{F}(\mathcal{M})$ is a Lie subalgebra of $(\mathscr{F}(\mathcal{M}),\{\cdot,\cdot\})$, making it a Poisson subalgebra. ■

► Remarks:

- If the quotient $\mathcal{M}/\mathbf{G}$ is a smooth manifold then we may identify $\mathscr{F}(\mathcal{M}/\mathbf{G})$ with $\mathscr{F}_{\mathbf{G}}(\mathcal{M})$ and Theorem 2.7 states that $\mathcal{M}/\mathbf{G}$ carries a Poisson structure for which the quotient map $\mathcal{M}\to\mathcal{M}/\mathbf{G}$ is a morphism of Poisson manifolds.
- The quotient structure usually admits Casimirs, so that we get Poisson or symplectic structures on the level sets of them. The case in which symplectic structures are found by the above procedure corresponds to what is classically known as *symplectic reduction*.

2.5 Lie-Poisson structures

► A fundamental example of Poisson structure is the so called *Lie-Poisson structure*, which is the linear degenerate Poisson structure associated with the dual of a finite-dimensional Lie algebra $\mathfrak{g}$. The identification of $\mathfrak{g}$ with $\mathfrak{g}^*$ by means of a symmetric non-degenerate bilinear Ad-invariant form (say the Killing form if $\mathfrak{g}$ is semi-simple or simple) allows then to define a Poisson structure directly on $\mathfrak{g}$. An example of Lie-Poisson structure has already been considered in Example 2.4.

2.5.1 *Lie-Poisson structure on* $\mathfrak{g}^*$

► Let $\mathfrak{g}$ be a n-dimensional real (or complex) Lie algebra, with Lie bracket $[\cdot,\cdot]$ and a non-degenerate Ad-invariant bilinear form $\langle\cdot\,|\,\cdot\rangle$, i.e., $\mathfrak{g}$ is a quadratic Lie algebra. Let $\mathfrak{g}^*$ be the dual of $\mathfrak{g}$. Recall that we denote by X any arbitrary element of $\mathfrak{g}$ and by X^* any arbitrary element of $\mathfrak{g}^*$.

Definition 2.6

Let $F, G \in \mathscr{F}(\mathfrak{g}^*)$. For any $\mathsf{X}^* \in \mathfrak{g}^*$ the **Lie-Poisson bracket** *(or* **Kostant-Kirillov bracket***) on $\mathfrak{g}^*$ is defined by*

$$\{F, G\}_{\mathrm{LP}}(\mathsf{X}^*) := \langle \mathsf{X}^*, [\, \mathsf{d}F(\mathsf{X}^*), \mathsf{d}G(\mathsf{X}^*)\,] \rangle . \tag{2.34}$$

We say that $\mathfrak{g}^*$ equipped with (2.34) is called **Lie-Poisson algebra** *and we denote by $(\mathfrak{g}^*, \{\cdot,\cdot\}_{\mathrm{LP}})$ the corresponding Poisson manifold.*

► Remarks:

- The Lie-Poisson structure (2.34) is induced by the Lie structure of $\mathfrak{g}$. Therefore its structure functions are linear functions with the structure constants of $\mathfrak{g}$ as coefficients.
- Skew-symmetry and bilinearity of (2.34) are obvious. The Jacobi identity is inherited by the Jacobi identity on $\mathfrak{g}$ (see below for a coordinate-dependent proof).
- The Poisson manifold $(\mathfrak{g}^*, \{\cdot,\cdot\}_{\mathrm{LP}})$ is always degenerate unless $\mathfrak{g}$ is Abelian.

► The next claim provides a form for Hamiltonian vector fields on $(\mathfrak{g}^*, \{\cdot,\cdot\}_{\mathrm{LP}})$.

Theorem 2.8

Let $\mathsf{X}^* \in \mathfrak{g}^*$ and $H \in \mathscr{F}(\mathfrak{g}^*)$. On $(\mathfrak{g}^*, \{\cdot,\cdot\}_{\mathrm{LP}})$ the Hamiltonian vector field v_H at X^* takes the form

$$\mathsf{v}_H(\mathsf{X}^*) = \mathsf{ad}^*_{\mathsf{d}H(\mathsf{X}^*)}\, \mathsf{X}^*. \tag{2.35}$$

Proof. Let $F \in \mathscr{F}(\mathfrak{g}^*)$. We have

$$\mathsf{v}_H(\mathsf{X}^*)[F] = \langle \mathsf{v}_H(\mathsf{X}^*), \mathsf{d}F(\mathsf{X}^*) \rangle .$$

On the other hand we have:

$$\begin{aligned} \mathsf{v}_H(\mathsf{X}^*)[F] &= \{F, H\}_{\mathrm{LP}}(\mathsf{X}^*) = \left\langle \mathsf{X}^*, -\mathsf{ad}_{\mathsf{d}H(\mathsf{X}^*)}\, \mathsf{d}F(\mathsf{X}^*) \right\rangle \\ &= \left\langle \mathsf{ad}^*_{\mathsf{d}H(\mathsf{X}^*)}\, \mathsf{X}^*, \mathsf{d}F(\mathsf{X}^*) \right\rangle , \end{aligned}$$

where we used (1.40). The claim follows. ■

► It is useful to provide some expressions in coordinates.

- Let $\{\mathsf{E}_1, \ldots, \mathsf{E}_n\}$ be an arbitrary basis on $\mathfrak{g}$. Denoting by c_{ij}^k the structure constants of $\mathfrak{g}$ in the basis $\{\mathsf{E}_1, \ldots, \mathsf{E}_n\}$, we have

$$[\, \mathsf{E}_i, \mathsf{E}_j \,] = \sum_{k=1}^{n} c_{ij}^k\, \mathsf{E}_k.$$

- Let $\{\mathsf{E}_1^*, \ldots, \mathsf{E}_n^*\}$ be the dual basis of $\mathfrak{g}^*$ with
$$\langle \mathsf{E}_i, \mathsf{E}_j^* \rangle = \delta_{ij}.$$
Then an element of $\mathfrak{g}^*$ can be represented as
$$\mathsf{X}^* = \sum_{i=1}^{n} x_i \, \mathsf{E}_i^*,$$
for some smooth coordinates x_i which form a functional basis of $\mathscr{F}(\mathfrak{g}^*)$.

- Let us write the bracket (2.34) in coordinates x_i. We have:
$$\langle \mathsf{X}^*, \mathsf{E}_j \rangle = \sum_{i=1}^{n} x_i \, \langle \mathsf{E}_i^*, \mathsf{E}_j \rangle = x_j,$$
from which there follows that $\mathrm{d}x_j(\mathsf{X}^*) = \mathsf{E}_j$. By using formula (2.34) we have:
$$\begin{aligned}
\{x_i, x_j\}_{\mathrm{LP}}(\mathsf{X}^*) &= \langle \mathsf{X}^*, [\, \mathrm{d}x_i(\mathsf{X}^*), \mathrm{d}x_j(\mathsf{X}^*) \,] \rangle \\
&= \sum_{k=1}^{n} x_k \, \langle \mathsf{E}_k^*, [\, \mathsf{E}_i, \mathsf{E}_j \,] \rangle \\
&= \sum_{k,\ell=1}^{n} c_{ij}^{\ell} \, x_k \, \langle \mathsf{E}_k^*, \mathsf{E}_\ell \rangle .
\end{aligned}$$
Therefore we can write
$$\{x_i, x_j\}_{\mathrm{LP}} = \sum_{k=1}^{n} c_{ij}^{k} \, x_k.$$
Note that, as expected, the structure functions of the Poisson structure are linear functions with the structure constants of $\mathfrak{g}$ as coefficients. This also shows that the bracket (2.34) satisfies the Jacobi identity.

- Let us write the Hamiltonian vector field (2.35) in coordinates x_i. First of all we observe that for $H \in \mathscr{F}(\mathfrak{g}^*)$ we have:
$$\mathrm{d}H(\mathsf{X}^*) = \sum_{i=1}^{n} \frac{\partial H}{\partial x_i} \, \mathrm{d}x_i(\mathsf{X}^*) = \sum_{i=1}^{n} \frac{\partial H}{\partial x_i} \, \mathsf{E}_i.$$
The Hamilton equations for x_i generated by the Hamiltonian H are:
$$\begin{aligned}
\{x_i, H\}_{\mathrm{LP}}(\mathsf{X}^*) &= \sum_{k=1}^{n} x_k \, \langle \mathsf{E}_k^*, [\, \mathrm{d}x_i(\mathsf{X}^*), \mathrm{d}H(\mathsf{X}^*) \,] \rangle \\
&= \sum_{k,\ell=1}^{n} x_k \frac{\partial H}{\partial x_\ell} \, \langle \mathsf{E}_k^*, [\, \mathsf{E}_i, \mathsf{E}_\ell \,] \rangle \\
&= \sum_{k,\ell,m=1}^{n} c_{i\ell}^{m} \, x_k \, \frac{\partial H}{\partial x_\ell} \, \langle \mathsf{E}_k^*, \mathsf{E}_m \rangle .
\end{aligned}$$

Therefore we can write

$$\dot{x}_i = \sum_{k,\ell=1}^{n} c_{i\ell}^{k}\, x_k \frac{\partial H}{\partial x_\ell}, \qquad i = 1, \ldots, n. \tag{2.36}$$

Example 2.14 (*The Lie-Poisson algebra $\mathfrak{e}(3)^*$ and the Kirchhoff rigid body*)

Consider the euclidean six-dimensional Lie algebra $\mathfrak{e}(3) \simeq \mathfrak{so}(3) \oplus_s \mathbb{R}^3$ (see Example 1.23). It is the Lie algebra of $\mathbf{E}(3)$, the Lie group of rigid motions on $\mathbb{R}^3$ (rotations and translations). Let us parametrize the space $\mathfrak{e}(3)^*$ with coordinates $x := (x_1, x_2, x_3)^\top, y := (y_1, y_2, y_3)^\top \in \mathbb{R}^3$.

- The Lie-Poisson algebra $\mathfrak{e}(3)^*$ is completely specified by the Lie-Poisson brackets

$$\{x_i, x_j\}_{\mathrm{LP}} = \varepsilon_{ijk}\, x_k, \qquad \{x_i, y_j\}_{\mathrm{LP}} = \varepsilon_{ijk}\, y_k, \qquad \{y_i, y_j\}_{\mathrm{LP}} = 0. \tag{2.37}$$

- Formula (2.37) implies that

$$\begin{aligned}\{F(x,y), G(x,y)\}_{\mathrm{LP}} &= \langle\, x, \operatorname{grad}_x F(x,y) \times \operatorname{grad}_x G(x,y)\,\rangle \\ &\quad + \Big\langle\, y, \operatorname{grad}_x F(x,y) \times \operatorname{grad}_y G(x,y)\,\Big\rangle \\ &\quad + \Big\langle\, y, \operatorname{grad}_y F(x,y) \times \operatorname{grad}_x G(x,y)\,\Big\rangle,\end{aligned}$$

  where $\langle\,\cdot,\cdot\,\rangle$ denotes the standard scalar product in $\mathbb{R}^3$ and $F, G \in \mathscr{F}(\mathbb{R}^6)$. This is exactly the Lie-Poisson bracket (2.34) for $\mathfrak{e}(3)^*$.

- The Poisson structure is degenerate. Precisely its rank is $2r = 4$. It becomes non-degenerate on the orbits of the coadjoint representation (see Theorem 2.11),

$$\mathcal{O}_{(x,y)} := \Big\{(x,y) \in \mathbb{R}^6 \,:\, \langle\, y, y\,\rangle = c^2, \langle\, x, y\,\rangle = d\Big\},$$

  where $c, d \in \mathbb{R}$ are constants.

- For $H \in \mathscr{F}(\mathbb{R}^6)$ the Hamiltonian dynamics on $\mathfrak{e}(3)^*$ can be written as

$$\begin{cases} \dot{x} = x \times \operatorname{grad}_x H(x,y) + y \times \operatorname{grad}_y H(x,y), \\ \dot{y} = y \times \operatorname{grad}_x H(x,y). \end{cases} \tag{2.38}$$

  This is just a specialization of (2.35).

- If we fix

$$H(x,y) := \frac{1}{2}\sum_{i=1}^{3}\Big(a_i\, x_i^2 + b_i\, y_i^2 + e_i\, x_i\, y_i\Big), \qquad a_i, b_i, e_i \in \mathbb{R},$$

  the Hamiltonian equations (2.38) turn into the *Kirchhoff rigid body equations* describing the motion of a rigid body in an ideal fluid:

$$\begin{cases} \dot{x}_1 = (a_2 - a_3)\, x_2\, x_3 + (b_2 - b_3)\, y_2\, y_3 + (e_2 - e_3)(x_2\, y_3 + x_3\, y_2), \\ \dot{x}_2 = (a_3 - a_1)\, x_3\, x_1 + (b_3 - b_1)\, y_3\, y_1 + (e_3 - e_1)(x_3\, y_1 + x_1\, y_3), \\ \dot{x}_3 = (a_1 - a_2)\, x_1\, x_2 + (b_1 - b_2)\, y_1\, y_2 + (e_1 - e_2)(x_1\, y_2 + x_2\, y_1), \\ \dot{y}_1 = (e_2 - e_3)\, y_2\, y_3 + a_2\, x_2\, y_3 - a_3 x_3\, y_2, \\ \dot{y}_2 = (e_3 - e_1)\, y_3\, y_1 + a_3\, x_3\, y_1 - a_1\, x_1\, y_3, \\ \dot{y}_3 = (e_1 - e_2)\, y_1\, y_2 + a_1\, x_1\, y_2 - a_2\, x_2\, y_1. \end{cases} \tag{2.39}$$

2.5.2 Lie-Poisson structure on $\mathfrak{g}$

► Recall that $\mathfrak{g}$ is a quadratic Lie algebra. By identifying $\mathfrak{g}$ with its dual we obtain a linear Poisson structure on $\mathfrak{g}$.

- There exists an isomorphism $\sigma : \mathfrak{g} \to \mathfrak{g}^*$ which assigns to $\mathsf{X} \in \mathfrak{g}$ the linear form $\sigma(\mathsf{X}) \in \mathfrak{g}^*$ according to (see (1.48))
$$\langle \sigma(\mathsf{X}), \mathsf{Y} \rangle = \langle \mathsf{X} \,|\, \mathsf{Y} \rangle \qquad \forall \mathsf{X}, \mathsf{Y} \in \mathfrak{g}.$$
- The inverse of σ is the linear map $\sigma^{-1} : \mathfrak{g}^* \to \mathfrak{g} : \mathsf{X}^* \mapsto \mathsf{X}$, where X is the unique element of $\mathfrak{g}$ that satisfies $\langle \mathsf{X} \,|\, \mathsf{Y} \rangle = \langle \mathsf{X}^*, \mathsf{Y} \rangle$ for all $\mathsf{Y} \in \mathfrak{g}$. These isomorphisms allow us to associate to a function on $\mathfrak{g}^*$ a function on $\mathfrak{g}$ and viceversa, simply by composing with σ or σ^{-1}. The maps σ, σ^{-1} allow us to identify $\mathfrak{g}$ with $\mathfrak{g}^*$.

► We now prove that σ induces a Poisson structure on $\mathfrak{g}$, which is called *Lie-Poisson structure on* $\mathfrak{g}$ and is exactly the Poisson structure corresponding to the Lie-Poisson structure on $\mathfrak{g}^*$.

Theorem 2.9

The Lie-Poisson structure on $\mathfrak{g}$ (w.r.t. $\langle \cdot \,|\, \cdot \rangle$) is given by
$$\{F, G\}_{\mathrm{LP}}(\mathsf{X}) = \left\langle \mathsf{X} \,\middle|\, \left[\sigma^{-1}(\mathsf{d}F(\mathsf{X})), \sigma^{-1}(\mathsf{d}G(\mathsf{X})) \right] \right\rangle, \qquad F, G \in \mathscr{F}(\mathfrak{g}). \tag{2.40}$$

Proof. We proceed by steps.

- Let $F, G \in \mathscr{F}(\mathfrak{g})$ and compute their Lie-Poisson bracket $\{F, G\}_{\mathrm{LP}}$ (w.r.t. $\langle \cdot \,|\, \cdot \rangle$) at $\mathsf{X} \in \mathfrak{g}$ by using (2.34):
$$\begin{aligned} \{F, G\}_{\mathrm{LP}}(\mathsf{X}) &= \left\{ (\sigma^{-1})^\star F, (\sigma^{-1})^\star G \right\}_{\mathrm{LP}} (\sigma(\mathsf{X})) \\ &= \left\langle \sigma(\mathsf{X}), \left[\mathsf{d}\left((\sigma^{-1})^\star F \right)(\sigma(\mathsf{X})), \mathsf{d}\left((\sigma^{-1})^\star G \right)(\sigma(\mathsf{X})) \right] \right\rangle. \end{aligned} \tag{2.41}$$
- Let us now compute explicitly $\mathsf{d}\left((\sigma^{-1})^\star F \right)(\sigma(\mathsf{X}))$, viewed as an element of $\mathfrak{g}$. First use the chain rule and the fact that σ^{-1} is a linear map:
$$\mathsf{d}\left((\sigma^{-1})^\star F \right)(\sigma(\mathsf{X})) = \mathsf{d}\left(F \circ \sigma^{-1} \right)(\sigma(\mathsf{X})) = \mathsf{d}F(\mathsf{X}) \circ \sigma^{-1},$$
so that, for any $\mathsf{X}^* \in \mathfrak{g}^*$,
$$\begin{aligned} \left\langle \mathsf{d}\left((\sigma^{-1})^\star F \right)(\sigma(\mathsf{X})), \mathsf{X}^* \right\rangle &= \left\langle \mathsf{d}F(\mathsf{X}), \sigma^{-1}(\mathsf{X}^*) \right\rangle \\ &= \left\langle \sigma^{-1}(\mathsf{d}F(\mathsf{X})) \,\middle|\, \sigma^{-1}(\mathsf{X}^*) \right\rangle \\ &= \left\langle \sigma^{-1}(\mathsf{d}F(\mathsf{X})), \mathsf{X}^* \right\rangle. \end{aligned}$$

- There follows that

$$\mathsf{d}\left(\left(\sigma^{-1}\right)^{\star}F\right)(\sigma(\mathsf{X}))=\sigma^{-1}(\mathsf{d}F(\mathsf{X})).$$

- Therefore, from (2.41) we get

$$\begin{aligned}\{F,G\}_{\mathrm{LP}}(\mathsf{X}) &= \left\langle \sigma(\mathsf{X}),\left[\sigma^{-1}(\mathsf{d}F(\mathsf{X})),\sigma^{-1}(\mathsf{d}G(\mathsf{X}))\right]\right\rangle \\ &= \left\langle \mathsf{X}\,\middle|\,\left[\sigma^{-1}(\mathsf{d}F(\mathsf{X})),\sigma^{-1}(\mathsf{d}G(\mathsf{X}))\right]\right\rangle,\end{aligned}$$

  which is the claim.

The Theorem is proved. ∎

► Remarks:

- Formula (2.40) is usually written in a more readable form by setting

$$\nabla F(\mathsf{X}):=\sigma^{-1}(\mathsf{d}F(\mathsf{X})),$$

  which means

$$\langle\,\nabla F(\mathsf{X})\,|\,\mathsf{Y}\,\rangle=\langle\,\mathsf{d}F(\mathsf{X}),\mathsf{Y}\,\rangle\,, \tag{2.42}$$

  for $F\in\mathscr{F}(\mathfrak{g})$, $\mathsf{X},\mathsf{Y}\in\mathfrak{g}$. Since F is a function on a vector space, the latter definition can also be written in the following form:

$$\langle\,\nabla F(\mathsf{X})\,|\,\mathsf{Y}\,\rangle:=\left.\frac{\mathsf{d}}{\mathsf{d}\varepsilon}\right|_{\varepsilon=0}F(\mathsf{X}+\varepsilon\,\mathsf{Y}).$$

► Hence, we now give a definition analog to Definition 2.6.

Definition 2.7

Let $F,G\in\mathscr{F}(\mathfrak{g})$. For any $\mathsf{X}\in\mathfrak{g}$ the **Lie-Poisson bracket** *on $\mathfrak{g}$ is defined by*

$$\{F,G\}_{\mathrm{LP}}(\mathsf{X}):=\langle\,\mathsf{X}\,|\,[\,\nabla F(\mathsf{X}),\nabla G(\mathsf{X})\,]\,\rangle\,. \tag{2.43}$$

We denote by $(\mathfrak{g},\{\,\cdot,\cdot\,\}_{\mathrm{LP}})$ the corresponding Poisson manifold.

► We now give the analog of Theorem 2.8 on $(\mathfrak{g},\{\,\cdot,\cdot\,\}_{\mathrm{LP}})$.

Theorem 2.10

Let $\mathsf{X}\in\mathfrak{g}$ and $H\in\mathscr{F}(\mathfrak{g})$. On $(\mathfrak{g},\{\,\cdot,\cdot\,\}_{\mathrm{LP}})$ the Hamiltonian vector field v_H at X

takes the form

$$\mathsf{v}_H(\mathsf{X}) = [\nabla H(\mathsf{X}), \mathsf{X}]. \tag{2.44}$$

Proof. In fact, (1.49) implies that (2.43), for $H = G$, can be written as

$$\begin{aligned}\{F, H\}_{\mathrm{LP}}(\mathsf{X}) &= \langle \mathsf{X} \,|\, [\nabla F(\mathsf{X}), \nabla H(\mathsf{X})] \rangle = \langle [\mathsf{X}, \nabla F(\mathsf{X})] \,|\, \nabla H(\mathsf{X}) \rangle \\ &= \langle [\nabla H(\mathsf{X}), \mathsf{X}] \,|\, \nabla F(\mathsf{X}) \rangle = \langle [\nabla H(\mathsf{X}), \mathsf{X}], \mathrm{d}F(\mathsf{X}) \rangle .\end{aligned}$$

If we compare this to

$$\{F, H\}_{\mathrm{LP}}(\mathsf{X}) := \mathsf{v}_H(\mathsf{X})[F] = \langle \mathsf{v}_H(\mathsf{X}), \mathrm{d}F(\mathsf{X}) \rangle ,$$

the claim follows. ■

2.5.3 *Symplectic foliation*

► Let $\mathbf{G}$ be the Lie group of $\mathfrak{g}$. Recall that if $\mathsf{X}^* \in \mathfrak{g}^*$ the coadjoint orbit through X^* is

$$\mathcal{O}_{\mathsf{X}^*} := \{\mathrm{Ad}_h^* \mathsf{X}^* \,:\, h \in \mathbf{G}\}.$$

We now prove that if $\mathbf{G}$ is connected then the coadjoint orbits of $\mathbf{G}$ are symplectic manifolds.

Theorem 2.11 (*Kostant-Kirillov*)

Let $\mathfrak{g}$ be a finite-dimensional Lie algebra and let $\mathbf{G}$ be the connected Lie group whose Lie algebra is $\mathfrak{g}$. The symplectic leaves of the Poisson manifold $(\mathfrak{g}^*, \{\cdot,\cdot\}_{\mathrm{LP}})$ are the coadjoint orbits of $\mathbf{G}$.

Proof. We proceed by steps.

- The fundamental vector fields of a group action span all tangent spaces to the orbits of the action. Hence, the vector fields $\mathrm{ad}^*_{\mathsf{X}}$ span the tangent space to the coadjoint orbits $\mathcal{O}_{\mathsf{X}^*}$ at any point of $\mathsf{X}^* \in \mathfrak{g}^*$.
- For any $\mathsf{X}^* \in \mathfrak{g}^*$ this tangent space is:

$$T_{\mathsf{X}^*}\mathcal{O}_{\mathsf{X}^*} = \{\mathrm{ad}^*_{\mathsf{X}} \mathsf{X}^* \,:\, \mathsf{X} \in \mathfrak{g}\}.$$

- Two points belong to the same symplectic leaf if they can be joined by a path whose tangent vector at any point is a Hamiltonian vector field. We have from (2.35) that

$$\begin{aligned}\mathrm{Ham}_{\mathsf{X}^*}(\mathfrak{g}^*) &= \left\{\mathrm{ad}^*_{\mathrm{d}H(\mathsf{X}^*)} \mathsf{X}^* \,:\, H \in \mathscr{F}(\mathfrak{g}^*)\right\} \\ &= \{\mathrm{ad}^*_{\mathsf{X}} \mathsf{X}^* \,:\, \mathsf{X} \in \mathfrak{g}\},\end{aligned}$$

  where we used that, given $\mathsf{X} \in \mathfrak{g}$, the linear function H on $\mathfrak{g}^*$, defined by $H(\mathsf{X}^*) = \langle \mathsf{X}, \mathsf{X}^* \rangle$ realizes $\mathrm{d}H(\mathsf{X}^*) = \mathsf{X}$. The claim follows.

The Theorem is proved. ■

► It can be proved that the 2-form defined by

$$\omega\left(\mathsf{ad}^*_{\mathsf{d}F(\mathsf{X}^*)}\mathsf{X}^*,\mathsf{ad}^*_{\mathsf{d}G(\mathsf{X}^*)}\mathsf{X}^*\right):=\langle\mathsf{X}^*,[\,\mathsf{d}F(\mathsf{X}^*),\mathsf{d}G(\mathsf{X}^*)\,]\,\rangle$$

is the symplectic 2-form of the coadjoint orbit $\mathscr{O}_{\mathsf{X}^*}$.

► We now make some important observations about the Casimirs of the Lie-Poisson structures $(\mathfrak{g}^*,\{\,\cdot,\cdot\,\}_{\mathrm{LP}})$ and $(\mathfrak{g},\{\,\cdot,\cdot\,\}_{\mathrm{LP}})$.

- The set of Ad^*-invariant functions on $\mathfrak{g}^*$ is precisely the set of Casimirs for the Lie-Poisson bracket (see Theorem 1.1). This is evident if we consider the formula

  $$\{F,G\}_{\mathrm{LP}}(\mathsf{X}^*)=\langle\mathsf{X}^*,[\,\mathsf{d}F(\mathsf{X}^*),\mathsf{d}G(\mathsf{X}^*)\,]\,\rangle=\left\langle\,\mathsf{ad}^*_{\mathsf{d}G(\mathsf{X}^*)}\mathsf{X}^*,\mathsf{d}F(\mathsf{X}^*)\,\right\rangle.$$

  Therefore we have that $G\in\mathscr{F}(\mathfrak{g}^*)$ is a Casimir of $(\mathfrak{g}^*,\{\,\cdot,\cdot\,\}_{\mathrm{LP}})$ if

  $$\mathsf{ad}^*_{\mathsf{d}G(\mathsf{X}^*)}\mathsf{X}^*=0\qquad\forall\mathsf{X}^*\in\mathfrak{g}^*.$$

- Recall that Ad-invariance of $\langle\,\cdot\,|\,\cdot\,\rangle$ implies (see (1.50)),

  $$\mathsf{Ad}^*_h\,\sigma(\mathsf{X})=\sigma(\mathsf{Ad}_h\mathsf{X}),\qquad\forall h\in\mathbf{G},\,\mathsf{X}\in\mathfrak{g}.$$

  This allows us to identify adjoint and coadjoint actions. In other words, the isomorphism σ establishes a one-to-one correspondence between Ad^*-invariant functions on $\mathfrak{g}^*$ and Ad-invariant functions on $\mathfrak{g}$. Since the Casimirs of the Lie-Poisson structure on $\mathfrak{g}^*$ are the Ad^*-invariant functions on $\mathfrak{g}^*$, there follows that the Casimirs of the Lie-Poisson structure on $\mathfrak{g}$ are the Ad-invariant functions on $\mathfrak{g}$. In view of this identification we have that $G\in\mathscr{F}(\mathfrak{g})$ is a Casimir of $(\mathfrak{g},\{\,\cdot,\cdot\,\}_{\mathrm{LP}})$ if

  $$[\,\nabla G(\mathsf{X}),\mathsf{X}\,]=0\qquad\forall\mathsf{X}\in\mathfrak{g}.\tag{2.45}$$

► We now prove the following claim.

Theorem 2.12

Let $\mathfrak{g}\subseteq\mathfrak{gl}(n,\mathbb{R})$ and v_H, $H\in\mathscr{F}(\mathfrak{g})$, be a Hamiltonian vector field on $(\mathfrak{g},\{\,\cdot,\cdot\,\}_{\mathrm{LP}})$, obeying (2.44) at $\mathsf{X}\in\mathfrak{g}$. Then all coefficients of the characteristic polynomial of X are in Poisson involution with H.

Proof. We proceed by steps.

- Let us define $\mathsf{Y} := \nabla H(\mathsf{X})$, which is a map $\mathsf{Y} : \mathfrak{g} \to \mathfrak{gl}(n, \mathbb{R})$. Both X and Y are functions of the time t. In particular, $\mathsf{X} = \mathsf{X}(t)$ is an integral curve $\mathcal{I}_\varepsilon \to \mathfrak{gl}(n, \mathbb{R})$ of (2.44) defined for t in a neighborhood $\mathcal{I}_\varepsilon$ of 0.
- Recall that each coefficient of the characteristic polynomial of X can be expressed in terms of traces of powers of X ("Newton identities").
- Therefore the claim is proved if we prove that

$$\frac{\mathsf{d}}{\mathsf{d}t} \mathsf{Trace}\, \mathsf{X}^k := \left\{ \mathsf{Trace}\, \mathsf{X}^k, H \right\}_{\mathrm{LP}} = 0 \qquad \forall k \in \mathbb{N}.$$

- Take any $k \in \mathbb{N}$ and use (2.44) and the properties of the trace:

$$\begin{aligned} \frac{\mathsf{d}}{\mathsf{d}t} \mathsf{Trace}\, \mathsf{X}^k &= k\, \mathsf{Trace} \left(\mathsf{X}^{k-1} \frac{\mathsf{d}}{\mathsf{d}t} \mathsf{X} \right) \\ &= k\, \mathsf{Trace} \left(\mathsf{X}^{k-1} (\mathsf{Y}\mathsf{X} - \mathsf{X}\mathsf{Y}) \right) \\ &= k\, \mathsf{Trace} \left(\mathsf{X}^{k-1} \mathsf{Y}\mathsf{X} - \mathsf{X}^k \mathsf{Y} \right) \\ &= k\, \mathsf{Trace} \left(\mathsf{X}^k \mathsf{Y} - \mathsf{X}^k \mathsf{Y} \right) = 0, \end{aligned}$$

  as desired.

The Theorem is proved. ■

Example 2.15 (*The Lie-Poisson algebra* $\mathfrak{gl}(n, \mathbb{R})$)

Consider the Lie algebra $\mathfrak{gl}(n, \mathbb{R})$ of all $n \times n$ matrices (see Example 1.25).

- A generic element of $\mathfrak{gl}(n, \mathbb{R})$ is written as

$$\mathsf{X} = \sum_{i,j=1}^{n} \mathsf{X}_{ij}\, \mathsf{E}_{ij}.$$

  where the $n \times n$ matrices E_{ij}, $i, j = 1, \ldots, n$, with matrix elements $(\mathsf{E}_{ij})_{k\ell} := \delta_{ik}\, \delta_{j\ell}$ form a basis for $\mathfrak{gl}(n, \mathbb{R})$. Here the smooth coordinates X_{ij} form a functional basis of $\mathscr{F}(\mathfrak{gl}(n, \mathbb{R}))$. Note that

$$\nabla \mathsf{X}_{ij}(\mathsf{X}) = \mathsf{E}_{ji}.$$

- The Lie brackets between the elements of the basis are

$$[\, \mathsf{E}_{ij}, \mathsf{E}_{k\ell} \,] = \delta_{jk}\, \mathsf{E}_{i\ell} - \delta_{\ell i}\, \mathsf{E}_{kj}.$$

- Instead of using the Killing form computed in Example 1.25 we use the the non-degenerate bilinear form

$$\langle\, \mathsf{X} \,|\, \mathsf{Y} \,\rangle = \mathsf{Trace}\, (\mathsf{X}\mathsf{Y}), \qquad \mathsf{X}, \mathsf{Y} \in \mathfrak{gl}(n, \mathbb{R}).$$

- The Lie-Poisson bracket is given by:

$$\begin{aligned}\{X_{ij}, X_{k\ell}\}_{LP}(X) &= \langle X \,|\, [\nabla X_{ij}(X), \nabla X_{k\ell}(X)] \rangle = \langle X \,|\, [E_{ji}, E_{\ell k}] \rangle \\ &= \sum_{m,s=1}^{n} X_{ms} \langle E_{ms} \,|\, \delta_{i\ell} E_{jk} - \delta_{kj} E_{\ell i} \rangle \\ &= \sum_{m,s=1}^{n} X_{ms} \left(\delta_{i\ell} \operatorname{Trace}(E_{ms} E_{jk}) - \delta_{kj} \operatorname{Trace}(E_{ms} E_{\ell i}) \right) \\ &= \sum_{m,s=1}^{n} X_{ms} \left(\delta_{i\ell} \delta_{mk} \delta_{sj} - \delta_{kj} \delta_{mi} \delta_{s\ell} \right) \\ &= \delta_{i\ell} X_{kj} - \delta_{kj} X_{i\ell}.\end{aligned}$$

- Let us show that for every $k \in \mathbb{N}$ the function $H_k : \mathfrak{gl}(n, \mathbb{R}) \to \mathbb{R}$ defined by

$$H_k(X) := \frac{1}{k} \operatorname{Trace} X^k$$

is a Casimir function of the Lie-Poisson structure. For $X, Y \in \mathfrak{g}$ we have

$$\langle \nabla H_k(X) \,|\, Y \rangle := \left. \frac{d}{d\varepsilon} \right|_{\varepsilon=0} H_k(X + \varepsilon Y) = \operatorname{Trace}\left(X^{k-1} Y \right) = \left\langle X^{k-1} \,\middle|\, Y \right\rangle,$$

from which there follows

$$[\nabla H_k(X), X] = \left[X^{k-1}, X \right] = 0 \qquad \forall X \in \mathfrak{gl}(n, \mathbb{R}).$$

► The modern theory of integrable systems began in 1967 with the observation, due to P. Lax, that it is sometimes possible to write a nonlinear equation in the so called *Lax form*:

$$\dot{X} = [Y, X], \tag{2.46}$$

where X and Y are linear operators that depend on dynamical variables.

- It is usually assumed that X contains all information on the initial data and Y is a function of X.

- A proof similar to the one of Theorem 2.12 shows that all coefficients of the characteristic polynomial of X, which are called *spectral invariants*, are integrals of motion for (2.46). For such a reason, the characteristic polynomial of X, $\det(X - \mu 1)$ is often called *generating function for the integrals of motion*.

- Since the r.h.s. of (2.46) contains only the commutator, one can try to rewrite it in a representation such that X and Y are elements of a Lie algebra.

► We now observe that our Hamiltonian equation (2.44) is just a particular instance of a Lax equation.

- However, in such a case the spectral invariants are Casimir functions of our Lie-Poisson bracket, and their conservation is a trivial fact which has nothing to do with the existence of actual integrals of motion. In other words, all spectral invariants considered in Theorem 2.12 give rise to a trivial dynamics.

- This disappointing result is a consequence of the fact that the Casimirs of $(\mathfrak{g}, \{\cdot,\cdot\}_{\mathrm{LP}})$ are defined in terms of (2.45).
- In spite of this initial drawback, the idea of using Lie-Poisson structures is of fundamental importance. Indeed, instead of the Lie-Poisson bracket (2.34), whose Casimirs are spectral invariants of matrices, one has to find some other Poisson structure. We will implement this idea in Chapter 4, after a proper definition of complete integrability.

Example 2.16 (*Lax equation for the Euler top*)

The system of ODEs of the Euler top (see Examples 2.4 and 2.12) can be put in a form of Lax equation in a very natural way. To to do we need a more physical description of the system.

- It is convenient to consider the equations of motion in a frame rotating with the body as discovered by Euler. We choose the moving frame with origin at the fixed point of the top (that is the point where the top is attached), and the axis being the principal inertia axis which diagonalize the inertia tensor computed with respect to the fixed point. Therefore, $I := \mathrm{diag}(I_1, I_2, I_3)$, where the parameters I_i are distinct and real. In the notation of Examples 2.4 and 2.12 we have $I_i = a_i^{-1}$.
- Let $x := (x_1, x_2, x_3)^\top \in \mathbb{R}^3$ be the angular momentum of the top in the moving frame. We have $x = I\,\omega$, where $\omega \in \mathbb{R}^3$ is the rotation vector of the moving frame.
- Equations of motion
$$\dot{x} = \omega \times x = \left(I^{-1}x\right) \times x \tag{2.47}$$
can be written in Hamiltonian form by defining the Poisson brackets of the Lie-Poisson algebra $\mathfrak{so}(3)^*$:
$$\{x_i, x_j\}_{\mathrm{LP}} = \varepsilon_{ijk} x_k. \tag{2.48}$$
- We know that the system admits two Poisson involutive and functional independent integrals of motion: the Hamiltonian
$$H(x) := \frac{1}{2}\left(\frac{x_1^2}{I_1} + \frac{x_2^2}{I_2} + \frac{x_3^2}{I_3}\right),$$
and the Casimir of (2.48):
$$B(x) := \frac{1}{2}\left(x_1^2 + x_2^2 + x_3^2\right).$$
- Thanks to the isomorphism $(\mathfrak{so}(3), [\cdot,\cdot]) \simeq (\mathbb{R}^3, \times)$ (see Example 1.22) we immediately notice that (2.47) is already a Lax equation. It suffices to define the 3×3 matrices X and Ω with entries respectively given by
$$\mathsf{X}_{ij} := \varepsilon_{ijk}\, x_k, \qquad \Omega_{ij} := \varepsilon_{ijk}\, \omega_k = \varepsilon_{ijk}\, \frac{x_k}{I_k},$$
to write (2.47) as
$$\dot{\mathsf{X}} = -[\Omega, \mathsf{X}].$$
This is a Lax equation on $\mathfrak{so}(3)$ and it is exactly of the form (2.44). Indeed we can write
$$\Omega_{ij} = \varepsilon_{ijk}\, \frac{\partial H}{\partial x_k}.$$

- If we now apply Theorem 2.12 we immediately notice that the coefficients of the characteristic polynomial of X give only the Casimir of the Lie-Poisson structure:

$$\operatorname{Trace} \mathsf{X}^{2k-1} = 0, \qquad \operatorname{Trace} \mathsf{X}^{2k} = 2\,(-1)^k\,(B(x))^{2k}, \qquad k \in \mathbb{N}.$$

2.5.4 Tensor formulation

► We now discuss the tensor formulation of the Lie-Poisson bracket on $\mathfrak{g}$:

$$\{F,G\}_{\mathrm{LP}}(\mathsf{X}) = \langle\, \mathsf{X} \,|\, [\, \nabla F(\mathsf{X}), \nabla G(\mathsf{X}) \,] \,\rangle\,,$$

with $F, G \in \mathscr{F}(\mathfrak{g})$.

- Let $\{\mathsf{E}_1, \ldots, \mathsf{E}_n\}$ be an arbitrary basis on $\mathfrak{g}$. Denoting by c_{ij}^k the structure constants of $\mathfrak{g}$ in the basis $\{\mathsf{E}_1, \ldots, \mathsf{E}_n\}$, we have

$$[\, \mathsf{E}_i, \mathsf{E}_j \,] = \sum_{k=1}^{n} c_{ij}^k\, \mathsf{E}_k.$$

- An element of $\mathfrak{g}$ can be represented as

$$\mathsf{X} = \sum_{i=1}^{n} \xi_i\, \mathsf{E}_i, \tag{2.49}$$

for some smooth coordinates ξ_i which form a functional basis of $\mathscr{F}(\mathfrak{g})$.

- Let $\{\mathsf{E}_1^*, \ldots, \mathsf{E}_n^*\}$ be the dual basis of $\mathfrak{g}^*$ with $\langle \mathsf{E}_i, \mathsf{E}_j^* \rangle = \delta_{ij}$. We have:

$$\left\langle\, \mathsf{X}, \mathsf{E}_j^* \,\right\rangle = \sum_{i=1}^{n} \xi_i \left\langle\, \mathsf{E}_i, \mathsf{E}_j^* \,\right\rangle = \xi_j,$$

from which there follows that $\nabla \xi_j(\mathsf{X}) = \mathsf{E}_j^*$.

- If A is the inverse of the $n \times n$ invertible matrix A^{-1} with entries $\langle \mathsf{E}_i \,|\, \mathsf{E}_j \rangle$ (*Killing matrix*), we can write

$$\mathsf{E}_i^* = \sum_{k=1}^{n} A_{ik}\, \mathsf{E}_k, \qquad A_{ik} = \langle \mathsf{E}_i^* \,|\, \mathsf{E}_k^* \rangle. \tag{2.50}$$

This formula encodes the identification of $\mathfrak{g}$ with $\mathfrak{g}^*$ by means of $\langle \cdot \,|\, \cdot \rangle$. Therefore if we write a Lie bracket $[\, \mathsf{E}_i^*, \mathsf{E}_j^* \,]$ between elements of $\mathfrak{g}^*$ this has to be interpreted according to (2.50) (see Example 1.24). Note also that

$$\langle \mathsf{E}_i^* \,|\, \mathsf{E}_j \rangle = \sum_{k=1}^{n} A_{ik} \langle \mathsf{E}_k \,|\, \mathsf{E}_j \rangle = \sum_{k=1}^{n} \langle \mathsf{E}_i^* \,|\, \mathsf{E}_k^* \rangle \langle \mathsf{E}_k \,|\, \mathsf{E}_j \rangle = \delta_{ij}.$$

► Now we give the following result.

Theorem 2.13

The coordinate representation of the Lie-Poisson bracket on $\mathfrak{g}$ takes the following form:

$$\{\,\xi_i,\xi_j\,\}_{\mathrm{LP}}(\mathsf{X}) = \frac{1}{2}\sum_{\ell,k=1}^{n} \xi_\ell \left(A_{ik}\, c^{j}_{\ell k} - A_{jk}\, c^{i}_{\ell k} \right). \tag{2.51}$$

Proof. We have:

$$\begin{aligned}
\{\,\xi_i,\xi_j\,\}_{\mathrm{LP}}(\mathsf{X}) &:= \langle \mathsf{X} \,|\, [\,\nabla\xi_i(\mathsf{X}),\nabla\xi_j(\mathsf{X})\,]\,\rangle = \langle \mathsf{X} \,|\, [\,\mathsf{E}_i^*,\mathsf{E}_j^*\,]\,\rangle \\
&= \frac{1}{2}\Big(\langle\, \mathsf{E}_j^* \,|\, [\mathsf{X},\mathsf{E}_i^*\,]\,\rangle - \langle\, \mathsf{E}_i^* \,|\, [\mathsf{X},\mathsf{E}_j^*\,]\,\rangle \Big),
\end{aligned}$$

where we used (1.49). Now, using (2.49) and (2.50) we get

$$\begin{aligned}
2\,\{\xi_i,\xi_j\}_{\mathrm{LP}}(\mathsf{X}) &= \sum_{\ell=1}^{n} \xi_\ell \,\langle\, \mathsf{E}_j^* \,|\, [\,\mathsf{E}_\ell,\mathsf{E}_i^*\,]\,\rangle - \sum_{\ell=1}^{n} \xi_\ell \,\langle\, \mathsf{E}_i^* \,|\, [\,\mathsf{E}_\ell,\mathsf{E}_j^*\,]\,\rangle \\
&= \sum_{\ell,k=1}^{n} \xi_\ell\, A_{ik} \,\langle\, \mathsf{E}_j^* \,|\, [\,\mathsf{E}_\ell,\mathsf{E}_k\,]\,\rangle - \sum_{\ell,k=1}^{n} \xi_\ell\, A_{jk} \,\langle\, \mathsf{E}_i^* \,|\, [\,\mathsf{E}_\ell,\mathsf{E}_k\,]\,\rangle \\
&= \sum_{\ell,k,r=1}^{n} \xi_\ell\, A_{ik}\, c^{r}_{\ell k} \,\langle\, \mathsf{E}_j^* \,|\, \mathsf{E}_r\,\rangle - \sum_{\ell,k,s=1}^{n} \xi_\ell\, A_{jk}\, c^{s}_{\ell k} \,\langle\, \mathsf{E}_i^* \,|\, \mathsf{E}_s\,\rangle \\
&= \sum_{\ell,k=1}^{n} \xi_\ell\, A_{ik}\, c^{j}_{\ell k} - \sum_{\ell,k=1}^{n} \xi_\ell\, A_{jk}\, c^{i}_{\ell k}.
\end{aligned}$$

The Theorem is proved. ■

► Let $\mathfrak{g}\otimes\mathfrak{g}$ be the Lie algebra obtained in terms of the tensor product of two copies of $\mathfrak{g}$. We define the following objects on $\mathfrak{g}\otimes\mathfrak{g}$:

$$\{\mathsf{X} \overset{\otimes}{,} \mathsf{X}\}_{\mathrm{LP}} := \sum_{i,j=1}^{n} \{\xi_i,\xi_j\}_{\mathrm{LP}}\, \mathsf{E}_i \otimes \mathsf{E}_j, \tag{2.52}$$

and

$$r := \frac{1}{2}\sum_{i,j=1}^{n} A_{ij}\, \mathsf{E}_i \otimes \mathsf{E}_j = \frac{1}{2}\sum_{i=1}^{n} \mathsf{E}_i \otimes \mathsf{E}_i^*. \tag{2.53}$$

To give a precise meaning to the above objects one has to embed $\mathfrak{g}$ into an associative algebra with unity 1.

► The next claim gives the tensor formulation of the Lie-Poisson bracket.

Theorem 2.14

The Lie-Poisson bracket on $\mathfrak{g}$ admits the following tensor formulation:

$$\{\mathsf{X} \overset{\otimes}{,} \mathsf{X}\}_{\mathrm{LP}} = [\,1 \otimes \mathsf{X} - \mathsf{X} \otimes 1, r\,]. \tag{2.54}$$

Proof. Let us compute the quantity $[\,1 \otimes \mathsf{X}, r\,]$ taking into account expression (2.53). We have:

$$\begin{aligned}
[\,1 \otimes \mathsf{X}, r\,] &= \frac{1}{2}\left[1 \otimes \sum_{\ell=1}^{n} \xi_\ell \, \mathsf{E}_\ell, \sum_{i,k=1}^{n} A_{ik} \, \mathsf{E}_i \otimes \mathsf{E}_k\right] \\
&= \frac{1}{2} \sum_{i,k,\ell=1}^{n} \xi_\ell \, A_{ik} \, \mathsf{E}_i \otimes [\,\mathsf{E}_\ell, \mathsf{E}_k\,] \\
&= \frac{1}{2} \sum_{i,j,k,\ell=1}^{n} \xi_\ell \, A_{ik} \, c^{j}_{\ell k} \, \mathsf{E}_i \otimes \mathsf{E}_j.
\end{aligned} \tag{2.55}$$

Similarly one finds:

$$[\,\mathsf{X} \otimes 1, r\,] = \frac{1}{2} \sum_{i,j,k,\ell=1}^{n} \xi_\ell \, A_{jk} \, c^{i}_{\ell k} \, \mathsf{E}_i \otimes \mathsf{E}_j. \tag{2.56}$$

Now, taking into account formula (2.52) and comparing the difference between (2.55) and (2.56) with (2.51) we get (2.54). ■

► Theorem 2.14 provides a convenient way to find Poisson submanifolds w.r.t. the Lie-Poisson bracket.

Theorem 2.15

Let $(\mathcal{M}, \{\,\cdot,\cdot\,\})$ be a Poisson manifold. Let $\mathsf{X} : \mathcal{M} \to \mathfrak{g}$ be a map such that the pairwise Poisson brackets of the coordinate functions ξ_j on $\mathcal{M}$ can be arranged as in the formula (2.54). The the image of the manifold $\mathcal{M}$ under the map X is a Poisson submanifold in $\mathfrak{g}$ equipped with the Lie-Poisson bracket.

Proof. The claim follows from the fact that (2.54) implies that X is a Poisson map if $\mathcal{M}$ is equipped with the Poisson bracket $\{\,\cdot,\cdot\,\}$ on $\mathcal{M}$ and $\mathfrak{g}$ is equipped with the Lie-Poisson bracket $\{\,\cdot,\cdot\,\}_{\mathrm{LP}}$ on $\mathfrak{g}$. ■

Example 2.17 (*The Lie-Poisson algebra $\mathfrak{gl}(n,\mathbb{R})$*)

Consider the Lie algebra $\mathfrak{gl}(n,\mathbb{R})$ of all $n \times n$ matrices (see Example 2.15), where a generic element is

$$\mathsf{X} = \sum_{i,j=1}^{n} \mathsf{X}_{ij} \, \mathsf{E}_{ij}.$$

- We already know that the Lie-Poisson algebra is

$$\{X_{ij}, X_{k\ell}\}_{\mathrm{LP}}(X) = \delta_{i\ell} X_{kj} - \delta_{kj} X_{i\ell}. \tag{2.57}$$

- It can be verified that the tensor formulation of (2.57) is given by

$$\{X \overset{\otimes}{,} X\}_{\mathrm{LP}} = [1 \otimes X - X \otimes 1, r],$$

with

$$r := \frac{1}{2} \sum_{j,k=1}^{N} E_{jk} \otimes E_{kj}.$$

3

Completely Integrable Systems

3.1 Introduction

► Completely integrable systems play a fundamental role in the realm of mathematical physics.

- Besides the fact that they offer a beautiful mathematical theory, they are the essential building blocks to construct perturbation theory of dynamical systems. In fact, loosely speaking, integrability of a dynamical system can be taken as a synonym of its solvability. Therefore, from its solutions one can construct in a perturbative way solutions of more complicated (non-integrable) systems.
- Liouville showed that the condition to have half of the dimension of phase space functionally independent and Poisson involutive integrals of motion is sufficient for integrating the equations of motion.
- It turns out that integrability of a Hamiltonian system does not only mean its solvability. A geometrical counterpart is that, under some topological assumptions, the motion of such a Hamiltonian system evolves on tori (the so called *Liouville tori*), whose dimension is half of the dimension of phase space, and the motion on them is quasi-periodic.

3.2 Definition of complete integrability

► In what follows we consider a real n-dimensional Poisson manifold $(\mathcal{M}, \{\cdot,\cdot\})$ with $\mathsf{rank}\{\cdot,\cdot\} := 2r \leqslant n$.

Definition 3.1

Let $\mathtt{F} := (F_1, \ldots, F_s)$, $F_i \in \mathscr{F}(\mathcal{M})$, $i = 1, \ldots, s$, be Poisson involutive and independent, with $\mathsf{s} := n - r$. Then we say that the set of Hamiltonian vector fields $\mathsf{v}_{F_1}, \ldots, \mathsf{v}_{F_s}$ defines a **completely integrable Hamiltonian system** *on $(\mathcal{M}, \{\cdot,\cdot\})$, which is denoted by the triple $(\mathcal{M}, \{\cdot,\cdot\}, \mathtt{F})$.*

► Remarks:

- If Definition 3.1 holds, the map $\mathtt{F} : \mathcal{M} \to \mathbb{R}^s$ is called the *momentum map* of $(\mathcal{M}, \{\cdot,\cdot\}, \mathtt{F})$. Note that among the s functions composing $\mathtt{F}$ there are Casimirs (which generate trivial dynamics) and actual integrals of motion (which generate non-trivial dynamics). Recall that the fiber of $\mathtt{F}$ that passes through $p \in \mathcal{M}$ is denoted by $\mathcal{F}_p$:

$$\mathcal{F}_p := \{\widetilde{p} \in \mathcal{M} \,:\, F_i(\widetilde{p}) = F_i(p),\ i = 1, \ldots, s\}.$$

- The integer r is called the *number of degrees of freedom* of $(\mathcal{M},\{\cdot,\cdot\},\mathtt{F})$. Notice that $2r$ is the dimension of the symplectic leaves of maximal dimension (typically the generic leaf) and that r is the number of independent commuting Hamiltonian vector fields on such a leaf.

- If $r=1$ integrability is trivial in the following sense. If $2r=2$ and there exist $n-2$ independent Casimirs $F_1,\ldots,F_{n-2}$, then, for a generic function F one has that $\mathtt{F}:=(F_1,\ldots,F_{n-2},F)$ is independent, so that $(\mathcal{M},\{\cdot,\cdot\},\mathtt{F})$ defines a trivial completely integrable system . Notice that the fibers of the momentum map $\mathtt{F}$ are in this case 1-dimensional.

- If $(\mathcal{M},\{\cdot,\cdot\})$ has maximal rank on $\mathcal{M}$, i.e., $2r=n$ (meaning that $\{\cdot,\cdot\}$ is non-degenerate), then $\mathcal{M}$ is a symplectic manifold and $(\mathcal{M},\{\cdot,\cdot\},\mathtt{F})$ is a completely integrable system if $\mathtt{F}:=(F_1,\ldots,F_r)$ is Poisson involutive and independent. In such a case one says that $(\mathcal{M},\{\cdot,\cdot\},\mathtt{F})$ is *completely canonically integrable.*

- Integrable systems often depend on parameters in the sense that one does not consider one particular Hamiltonian, but a parametric family of them. When the integrable systems come from physics these parameters usually have a physical meaning, such as moment of inertia, mass, spring constant, and so on. It is then understood that all claims and notions, such as independence of the functions, integrability, and so on, are valid for generic values of the parameters.

Example 3.1 (*n-dimensional harmonic oscillator*)

Consider the Hamiltonian phase space $\mathbb{R}^{2n}$, with canonical coordinates $(q,p):=(q_1,\ldots,q_n,p_1,\ldots,p_n)$ and with the Poisson structure coming from the canonical symplectic structure.

- The *n-dimensional harmonic oscillator* is defined by the Hamiltonian

$$H(q,p):=\frac{1}{2}\sum_{i=1}^{n}\left(p_i^2+\nu_i q_i^2\right),\qquad \nu_i>0,\ i=1,\ldots,n.$$

- It is easy to check that $\mathtt{F}:=(F_1,\ldots,F_n)$, with

$$F_i(q,p):=\frac{1}{2}\left(p_i^2+\nu_i q_i^2\right),\qquad i=1,\ldots,n,$$

is Poisson involutive and independent. Notice that $H=F_1+\cdots+F_n$, so that we may replace F_1 by H. We conclude that $(\mathbb{R}^{2n},\{\cdot,\cdot\},\mathtt{F})$ is a completely integrable system.

- Note that the fibers of the momentum map $\mathtt{F}$ over $(c_1,\ldots,c_n)$, with $c_i>0$, are products of circles $p_i^2+\nu_i q_i^2=c_i$. Hence they are n-dimensional tori.

- If $\nu_i=\nu_j$ for all $i,j=1,\ldots,n$, the system reduces to an *isotropic oscillator*. Notice that in this case each of the functions $q_i\,p_j-q_j\,p_i$ is an integral of motion of motion, but these functions are not all in involution.

Example 3.2 (*The Euler top*)

The Euler top is a classical example of a completely integrable system defined on the Poisson manifold

$\mathfrak{so}(3)^*$ (see Examples 2.4 and 2.12). In such a case the dimension of the Poisson manifold is $n = 3$, while its rank is $2r = 2$. The necessary number of Poisson involutive and functionally independent integrals of motion is $s = n - r = 3 - 1 = 2$, which is exactly the case. In the terminology of Definition 3.1 we can write $\mathtt{F} := (F_1, F_2)$ where

$$F_1(x) = \frac{1}{2}\left(a_1\, x_1^2 + a_2\, x_2^2 + a_3\, x_3^2\right), \qquad F_2(x) := \frac{1}{2}\left(x_1^2 + x_2^2 + x_3^2\right).$$

Example 3.3 (*The Lagrange top*)

The Lagrange top is a famous case of the rigid body motion around a fixed point in a constant gravitational field $(0, 0, \gamma)$. It is characterized by the following data: the inertia tensor is given by $\mathrm{diag}(1, 1, \alpha)$, $\alpha \in \mathbb{R}$, which means that the body is rotationally symmetric w.r.t. the third coordinate axis, and the fixed point lies on the symmetry axis.

- Equations of motion of the Lagrange top are of Kirchhoff type (see (2.39)), so that they are Hamiltonian w.r.t. the six-dimensional Lie-Poisson algebra $\mathfrak{e}(3)^*$:

$$\{x_i, x_j\}_{\mathrm{LP}} = \varepsilon_{ijk}\, x_k, \qquad \{x_i, y_j\}_{\mathrm{LP}} = \varepsilon_{ijk}\, y_k, \qquad \{y_i, y_j\}_{\mathrm{LP}} = 0,$$

  whose rank is $2r = 4$ due to the existence of the Casimirs

$$F_1(x, y) := \frac{1}{2}\left(y_1^2 + y_2^2 + y_3^2\right), \qquad F_2(x, y) := x_1\, y_1 + x_2\, y_2 + x_3\, y_3.$$

- The Hamiltonian reads

$$H(x, y) := \frac{1}{2}\left(x_1^2 + x_2^2 + x_3^2\right) + \gamma\, y_3,$$

  which is a special case of (**??**), and the explicit form of the equations of motion is

$$\begin{cases} \dot{x}_1 = (\alpha - 1)\, x_2\, x_3 + \gamma\, y_2, \\ \dot{x}_2 = (1 - \alpha)\, x_1\, x_3 - \gamma\, y_1, \\ \dot{x}_3 = 0, \\ \dot{y}_1 = \alpha\, y_2\, x_3 - y_3\, x_2, \\ \dot{y}_2 = y_3\, x_1 - \alpha\, y_1\, x_3, \\ \dot{y}_3 = y_1\, x_2 - y_2\, x_1. \end{cases}$$

- Note that we have three functionally independent and Poisson involutive integrals of motion. To have complete integrability of the system we need $s = n - r = 6 - 2 = 4$ functionally independent and Poisson involutive integrals of motion. Indeed, the system is completely integrable due to the existence of a fourth independent integral of motion, which is

$$F_3(x, y) := x_3,$$

  the third component of the angular momentum, as the rotational symmetry of the problem suggests. In the terminology of Definition 3.1 we can write $\mathtt{F} := (F_1, F_2, F_3, H)$.

Example 3.4 (*A one degree of freedom n-dimensional system*)

Consider the following set of quadratic differential equations in $\mathbb{R}^n$:

$$\dot{x}_i = x_i \sum_{j=1}^{n} x_j - \sum_{j=1}^{n} x_j^2, \qquad i = 1, \ldots, n. \tag{3.1}$$

- System (3.1) is a Hamiltonian system with Hamiltonian

$$H(x) = \frac{1}{2} \sum_{j=1}^{n} x_j^2$$

w.r.t. the linear Poisson structure defined by

$$\{ x_i, x_j \} = x_i - x_j. \qquad i, j = 1, \ldots, n, \tag{3.2}$$

that is $\dot{x}_i = \{ x_i, H(x) \}$.

- The Poisson structure (3.2) is degenerate with rank $2r = 2$. It has the functions

$$F_{ijk\ell}(x) = \frac{x_i - x_j}{x_k - x_\ell},$$

with $i \neq j, k \neq \ell$, as Casimir functions.

- Among the functions $F_{ijk\ell}$ only $n-2$ are functionally independent. Furthermore these $n-2$ functions are functionally independent on H, so that system (3.1) admits $n-1$ functionally independent integrals of motion.

3.2.1 Bi-Hamiltonian systems

► One of the most characteristic features of integrable systems is the existence of bi-Hamiltonian (or multi-Hamiltonian) structures. This ingenious concept was introduced by Magri in 1978. From the geometrical point of view, this means that there exists a pair of compatible Poisson 2-vector fields which allow us, using a recursion chain, to generate hierarchies of commuting symmetries and integrals of motion being in involution w.r.t. the above Poisson structures.

► We now prove that bi-Hamiltonian systems are good candidates to be completely integrable. Indeed, if a dynamical system admits two (or more) Hamiltonian compatible formulations then it automatically possesses a large number of integrals in involution. Nevertheless their independence is not guaranteed and it has to be checked a posteriori.

► We consider a bi-Hamiltonian manifold $(\mathcal{M}, \{\cdot,\cdot\}_1, \{\cdot,\cdot\}_2)$ where $\{\cdot,\cdot\}_1$ and $\{\cdot,\cdot\}_2$ are two compatible Poisson structures on $\mathcal{M}$. The following Theorem holds locally.

Theorem 3.1

Let $F_1, F_2 \in \mathscr{F}(\mathcal{M})$ and let $\mathrm{v}_{F_1} := \{\cdot, F_1\}_1$ and $\mathrm{v}_{F_2} := \{\cdot, F_2\}_2$ be the Hamiltonian vector fields generated by F_1 w.r.t. $\{\cdot,\cdot\}_1$ and by F_2 w.r.t. $\{\cdot,\cdot\}_2$. Denote by Φ_t the flow of the vector field generated by F_2 w.r.t. $\{\cdot,\cdot\}_1$.

1. *If*

$$\{ \cdot, F_1 \}_1 = \{ \cdot, F_2 \}_2, \tag{3.3}$$

then Φ_t consists of Poisson maps also w.r.t. $\{\cdot,\cdot\}_2$.

2. *Assume that $\{\cdot,\cdot\}_2$ has maximal rank and that condition (3.3) holds true. Then there exists a sequence of smooth functions $\mathrm{F} := (F_k)_{k=1}^{\infty}$, starting with F_1, F_2, such that:*

(a) $\{ \cdot, F_k \}_1 = \{ \cdot, F_{k+1} \}_2$ *for all* $k \geqslant 1$.

(b) $\mathtt{F}$ *is Poisson involutive w.r.t. both Poisson brackets:*

$$\{ F_k, F_j \}_1 = 0, \qquad \{ F_k, F_j \}_2 = 0 \qquad \forall\, j,k \geqslant 1.$$

Proof. We prove both claims.

1. For any two functions $H, G \in \mathscr{F}(\mathcal{M})$ define the function

$$K := \{ \Phi_t^\star H, \Phi_t^\star G \}_2 - \Phi_t^\star \{ H, G \}_2.$$

We want to prove that $K = 0$ for all $t \in \mathbb{R}$. We have:

$$\begin{aligned}\frac{\mathrm{d}}{\mathrm{d}t}K &= \left\{ \frac{\mathrm{d}}{\mathrm{d}t}(\Phi_t^\star H), \Phi_t^\star G \right\}_2 + \left\{ \Phi_t^\star H, \frac{\mathrm{d}}{\mathrm{d}t}(\Phi_t^\star G) \right\}_2 - \frac{\mathrm{d}}{\mathrm{d}t}(\Phi_t^\star \{ H, G \}_2) \\ &= \{ \{ \Phi_t^\star H, F_2 \}_1, \Phi_t^\star G \}_2 + \{ \Phi_t^\star H, \{ \Phi_t^\star G, F_2 \}_1 \}_2 \\ &\quad - \{ \Phi_t^\star \{ H, G \}_2, F_2 \}_1.\end{aligned}$$

Using now the Schouten-Nijenhuis bracket (2.3), formula (3.3) and the Jacobi identity for the bracket $\{ \cdot, \cdot \}_1$ a straightforward computation gives

$$\frac{\mathrm{d}}{\mathrm{d}t}K = \{ K, F_2 \}_1,$$

whose unique solution is $K(t) = K(0) \circ \Phi_t$. But $K(0) = 0$, so that $K = 0$ for all $t \in \mathbb{R}$.

2. We prove both statements.

(a) We know that the flow of a vector field on a symplectic manifold consists of symplectic morphisms if and only if this field is locally Hamiltonian. Therefore there exists at least locally a function $F_3 \in \mathscr{F}(\mathcal{M})$ such that

$$\{ \cdot, F_2 \}_1 = \{ \cdot, F_3 \}_2.$$

The construction can be iterated to give a sequence of smooth functions $\mathtt{F} := (F_k)_{k=1}^{\infty}$ starting with F_1, F_2 such that $\{ \cdot, F_k \}_1 = \{ \cdot, F_{k+1} \}_2$ for all $k \geqslant 1$.

(b) Let us prove that $\{ F_k, F_j \}_1 = 0$. We may assume that $k < j$. The property $\{ \cdot, F_k \}_1 = \{ \cdot, F_{k+1} \}_2$ for all $k \geqslant 1$ gives

$$\{ F_k, F_j \}_1 = \{ F_{k+1}, F_j \}_2 = \{ F_{k+1}, F_{j-1} \}_1.$$

By induction, for any $i \geqslant 1$, we get:

$$\{ F_k, F_j \}_1 = \{ F_{k+i}, F_{j-i+1} \}_2 = \{ F_{k+i}, F_{j-i} \}_1.$$

Now, assume that $j - k$ is an odd number. Then the middle term vanishes if $k + i = j - i + 1$, i.e., if $2i - 1 = j - k$. If we assume that $j - k$ is an even number we see that the last term vanishes if $k + i = j - i$, i.e., if $2i = j - k$.

The Theorem is proved. ■

► A drawback of the above construction is that not all functions of the sequence $F := (F_k)_{k=1}^{\infty}$ can be functionally independent (only a finite number of then can be). It might happen that the number s of independent functions generated by this construction is less than necessary to assure the complete integrability. However, in all known examples bi-Hamiltonian systems turn out to be completely integrable!

Example 3.5 (*A tri-Hamiltonian system on* $\mathbb{R}^6$)

In $\mathbb{R}^6$ consider the vector fields v_1 and v_2 corresponding respectively to the systems of ODEs

$$\begin{cases} \dot{x}_1 = 2\,x_5\,x_6, \\ \dot{x}_2 = 2\,x_3\,x_4, \\ \dot{x}_3 = x_5\,(x_1 + x_4), \\ \dot{x}_4 = 2\,x_2\,x_3, \\ \dot{x}_5 = x_3\,(x_1 + x_4), \\ \dot{x}_6 = 2\,x_5\,x_1. \end{cases} \qquad \begin{cases} \dot{x}_1 = x_2\,x_6, \\ \dot{x}_2 = x_4\,(2\,x_3 - x_6), \\ \dot{x}_3 = x_4\,x_5, \\ \dot{x}_4 = x_2(2\,x_3 - x_6), \\ \dot{x}_5 = x_3\,x_4, \\ \dot{x}_6 = x_1\,x_2. \end{cases}$$

- One can check that v_1 and v_2 admit four quadratic functionally independent integrals of motion, given by the functions

$$\begin{aligned} F_1(x) &= x_3^2 - x_5^2, \\ F_2(x) &= x_1^2 - x_6^2, \\ F_3(x) &= x_2^2 - x_4^2, \\ F_4(x) &= (x_1 - x_4)^2 - 2(x_2 - x_5)^2 - 2\,(x_3 - x_6)^2. \end{aligned}$$

- One can verify that there exist three linearly independent linear Poisson structures on $\mathbb{R}^6$, with rank four, with respect to which v_1 and v_2 are Hamiltonian. These Poisson structures are compatible, implying that the system is a completely integrable system with a tri-Hamiltonian structure.

- Explicitly, for $(\alpha, \beta, \gamma) \in \mathbb{R}^3$, consider the parametric skew-symmetric 6×6 matrix $\mathcal{X}_{\alpha\beta\gamma}$ whose non-vanishing entries are:

$$\begin{aligned} &(\mathcal{X}_{\alpha\beta\gamma})_{12} := \alpha\,x_6, \quad (\mathcal{X}_{\alpha\beta\gamma})_{13} := -\beta\,x_5, \quad (\mathcal{X}_{\alpha\beta\gamma})_{15} := -\beta\,x_3 - 2\,\gamma\,x_6, \\ &(\mathcal{X}_{\alpha\beta\gamma})_{16} := \beta(x_2 - 2\,x_5), \quad (\mathcal{X}_{\alpha\beta\gamma})_{23} := 2\,\gamma\,x_4, \quad (\mathcal{X}_{\alpha\beta\gamma})_{24} := \alpha(x_6 - 2\,x_3), \\ &(\mathcal{X}_{\alpha\beta\gamma})_{26} := -\alpha\,x_1 - \beta\,x_4, \quad (\mathcal{X}_{\alpha\beta\gamma})_{34} := -\alpha\,x_5 - 2\,\gamma\,x_2, \quad (\mathcal{X}_{\alpha\beta\gamma})_{35} := -\gamma(x_1 + x_4), \\ &(\mathcal{X}_{\alpha\beta\gamma})_{45} := \alpha\,x_3, \quad (\mathcal{X}_{\alpha\beta\gamma})_{46} := -\beta\,x_2, \quad (\mathcal{X}_{\alpha\beta\gamma})_{56} := 2\,\gamma\,x_1. \end{aligned}$$

For any $(\alpha, \beta, \gamma) \in \mathbb{R}^3$ $\mathcal{X}_{\alpha\beta\gamma}$ is the Poisson matrix of a Poisson structure on $\mathbb{R}^6$. $\mathcal{X}_{\alpha\beta\gamma}$ generates the Hamiltonian vector fields v_1 and v_2 according to the following *Lenard-Magri chains*. Setting $(\alpha, \beta, \gamma) = (1, 0, 0)$ we get

$$\begin{aligned} \mathcal{X}_{100}\ \mathrm{grad}_x F_1(x) &= 0, \\ \mathcal{X}_{100}\ \mathrm{grad}_x F_2(x) &= 0, \\ \mathcal{X}_{100}\ \mathrm{grad}_x F_3(x) &= 2\,v_2, \\ \mathcal{X}_{100}\ \mathrm{grad}_x F_4(x) &= 2\,(v_1 - 2\,v_2), \end{aligned}$$

so that F_1, F_2 are Casimirs of $\mathcal{X}_{100}$. Setting $(\alpha, \beta, \gamma) = (0, 1, 0)$ we get

$$\begin{aligned}
\mathcal{X}_{010}\ \mathsf{grad}_x F_1(x) &= 0,\\
\mathcal{X}_{010}\ \mathsf{grad}_x F_2(x) &= 2\,(\mathsf{v}_1 - \mathsf{v}_2),\\
\mathcal{X}_{010}\ \mathsf{grad}_x F_3(x) &= 0,\\
\mathcal{X}_{010}\ \mathsf{grad}_x F_4(x) &= 2\,(\mathsf{v}_1 - 2\,\mathsf{v}_2),
\end{aligned}$$

so that F_1, F_3 are Casimirs of $\mathcal{X}_{010}$. Setting $(\alpha, \beta, \gamma) = (0, 0, 1)$ we get

$$\begin{aligned}
\mathcal{X}_{001}\ \mathsf{grad}_x F_1(x) &= 2\,\mathsf{v}_1,\\
\mathcal{X}_{001}\ \mathsf{grad}_x F_2(x) &= 0,\\
\mathcal{X}_{001}\ \mathsf{grad}_x F_3(x) &= 0,\\
\mathcal{X}_{001}\ \mathsf{grad}_x F_4(x) &= 4\,(\mathsf{v}_1 - 2\,\mathsf{v}_2),
\end{aligned}$$

so that F_2, F_3 are Casimirs of $\mathcal{X}_{001}$.

3.2.2 Solvability by quadratures

► As anticipated the term *integrable system* reminds the fact that the system of ODEs representing it is solvable. On the other hand solvability of a dynamical system is a notion which is more general than complete integrability. In other words, a dynamical system can be solvable but not completely integrable (at least according to Definition 3.1).

Example 3.6 (*A solvable system which is not completely integrable*)

In $\mathbb{R}^3$ consider the following system of ODEs (*Ramanujan system*)

$$\begin{cases}
\dot{x}_1 = \dfrac{1}{12}\left(x_2 - x_1^2\right),\\[2mm]
\dot{x}_2 = \dfrac{1}{3}\left(x_3 - x_1\,x_2\right),\\[2mm]
\dot{x}_3 = \dfrac{1}{2}\left(x_2^2 - x_1\,x_3\right).
\end{cases}$$

The above system is known to be explicitly solvable but it is not completely integrable in the sense of Definition 3.1.

► We start with an example with one degree of freedom and we show how the equations of motion can be integrated explicitly, by using only algebraic operations, the process of taking inverse functions and integration.

Example 3.7 (*A particle moving in* $\mathbb{R}$ *under the influence of a potential energy*)

We consider a Hamiltonian system on the canonical phase space $\mathbb{R}^2$ defined by the Hamiltonian

$$H(q, p) := \frac{p^2}{2} + U(q). \tag{3.4}$$

The corresponding Hamilton equations describe the motion of a point-like particle with unit mass moving on the line $\mathbb{R}$ under the influence of a smooth potential energy U.

- We fix a point $(\widetilde{q},\widetilde{p}) \in \mathbb{R}^2$ for which $\mathrm{d}H(\widetilde{q},\widetilde{p}) \neq 0$ and we denote the value of H at $(\widetilde{q},\widetilde{p})$ by h. Using $\dot{q} = p$, the Hamiltonian (3.4) implies that the integral curve which starts at $(\widetilde{q},\widetilde{p})$ satisfies the differential equation

$$\mathrm{d}t = \frac{\mathrm{d}q}{\sqrt{2(h - U(q))}}. \tag{3.5}$$

The above denominator does not vanish in a neighborhood of $\widetilde{q}$, except maybe at $\widetilde{q}$, because $\mathrm{d}H(\widetilde{q},\widetilde{p}) \neq 0$.

- Integrating both sides of (3.5) we find ($t_0 = 0$):

$$t = \int_{\widetilde{q}}^{q(t)} \frac{\mathrm{d}\xi}{\sqrt{2(h - U(\xi))}},$$

which defines q (and hence also p) implicitly as a function of t. The obtained functions $(q(t), p(t))$ define, for $|t|$ small, the integral curve of the canonical Hamiltonian vector field v_H, starting at $(\widetilde{q},\widetilde{p})$, hence they integrate the equations of motion for the initial condition $(q(0), p(0)) = (\widetilde{q},\widetilde{p})$. Notice that q is obtained by using only algebraic operations, inverting a function and integration.

- A simple cases where the integration can be carried out in a closed form is the following: let $U(q) := \alpha^2 q^{-2}$, $\alpha \in \mathbb{R}, h > 0$. The motion takes place in the region $\widetilde{q} < q < \infty$, where $\widetilde{q} := \sqrt{\alpha^2/h}$. One finds:

$$q(t) = \sqrt{\widetilde{q}^2 + \widetilde{p}^2\, t^2}, \qquad \widetilde{p}^2 := 2\,h.$$

► We now show how the explicit integration of a completely integrable system can be done locally by using only algebraic operations, the "Inverse function Theorem" and integration. This process is usually called *solvability by quadratures*.

- Let $(\mathcal{M}, \{\cdot,\cdot\}, \mathrm{F})$ be a completely integrable system of rank $2r$ (recall that $s := n - r$).

- Let $p \in \mathcal{U}_{\mathrm{F}} \cap \mathcal{M}_{(r)}$, where

$$\mathcal{U}_{\mathrm{F}} := \{\widetilde{p} \in \mathcal{M} \,:\, \mathrm{d}F_1(\widetilde{p}) \wedge \cdots \wedge \mathrm{d}F_s(\widetilde{p}) \neq 0\},$$

and

$$\mathcal{M}_{(r)} := \left\{\widetilde{p} \in \mathcal{M} \,:\, \mathsf{rank}_{\widetilde{p}}\{\cdot,\cdot\} \geqslant 2r\right\}.$$

Note that the condition $p \in \mathcal{U}_{\mathrm{F}} \cap \mathcal{M}_{(r)}$ corresponds to the condition $\mathrm{d}H(q_0, p_0) \neq 0$ in Example 3.7.

- Recall that (see the third claim of Theorem 2.1):

$$\mathsf{dim}\left(\mathsf{span}\{\mathsf{v}_{F_1}(p), \ldots, \mathsf{v}_{F_s}(p)\}\right) = r, \tag{3.6}$$

for $p \in \mathcal{U}_{\mathrm{F}} \cap \mathcal{M}_{(r)}$.

- We want to show that the integral curve, starting at p of each of the Hamiltonian vector fields v_{F_i} can be obtained locally.

- In view of (3.6) we may suppose that the functions $F_1, \dots, F_s$ composing $\mathtt{F}$ are ordered in such a way that $\mathsf{v}_{F_1}, \dots, \mathsf{v}_{F_r}$ are independent at p. Then these vector fields are independent on an open neighborhood $\mathcal{V} \subset \mathcal{U}_{\mathtt{F}} \cap \mathcal{M}_{(r)}$ of p. Recall that $\mathcal{U}_{\mathtt{F}} \cap \mathcal{M}_{(r)}$ is open and contains p.

- Define $\mathcal{U} := \mathcal{V} \cap \mathcal{F}_p$. Recall that $\mathcal{F}_p$ is r-dimensional in view of (3.6).

- Since the vector fields $\mathsf{v}_{F_1}, \dots, \mathsf{v}_{F_r}$ are independent at every point of $\mathcal{U}$ there exist unique 1-forms $\omega_1, \dots, \omega_r \in \Omega(\mathcal{U})$ such that

$$\omega_i(\mathsf{v}_{F_j}) = \delta_{ij}, \qquad i, j = 1, \dots, r.$$

- The 1-forms ω_i can be computed as follows. For $G \in \mathscr{F}(\mathcal{U})$ we have

$$\mathsf{d}G = \sum_{i=1}^{r} \{ G, F_i \} \, \omega_i.$$

Proof. Evaluate both sides on the vector fields $\mathsf{v}_{F_1}, \dots, \mathsf{v}_{F_r}$, which span $T_p\mathcal{U}$ for any $p \in \mathcal{U}$:

$$\mathsf{d}G(\mathsf{v}_{F_j}) = \mathsf{v}_{F_j}[G] = \{ G, F_j \},$$

but

$$\sum_{i=1}^{r} \{ G, F_i \} \, \omega_i(\mathsf{v}_{F_j}) = \sum_{i=1}^{r} \{ G, F_i \} \, \delta_{ij} = \{ G, F_j \}.$$

- For $i = 1, \dots, r$, the 1-form ω_i is closed.

Proof. The vector fields $\mathsf{v}_{F_1}, \dots, \mathsf{v}_{F_r}$ span $T_p\mathcal{U}$ at every $p \in \mathcal{U}$ and (see (1.17)),

$$\mathsf{d}\omega_i \left(\mathsf{v}_{F_j}, \mathsf{v}_{F_k} \right) = \mathsf{v}_{F_j} \left[\omega_i \left(\mathsf{v}_{F_k} \right) \right] - \mathsf{v}_{F_k} \left[\omega_i \left(\mathsf{v}_{F_j} \right) \right] - \omega_i \left(\left[\mathsf{v}_{F_j}, \mathsf{v}_{F_k} \right] \right),$$

for any $j, k = 1, \dots, r$, which evaluates to zero because $\omega_i(\mathsf{v}_{F_j}) = \delta_{ij}$ and because the vector fields v_{F_j} commute.

- Now choose r functions $y_1, \dots, y_r$ on $\mathcal{U}$ whose differentials are independent at every point $p \in \mathcal{U}$, i.e., choose a system of coordinates on $\mathcal{U}$ (these functions may for example be chosen as r of the elements of a system of coordinates of $\mathcal{M}$ around p). Then,

$$\begin{pmatrix} \mathsf{d}y_1 \\ \vdots \\ \mathsf{d}y_r \end{pmatrix} = \begin{pmatrix} \{ y_1, F_1 \} & \cdots & \{ y_1, F_r \} \\ \vdots & & \vdots \\ \{ y_r, F_1 \} & \cdots & \{ y_r, F_r \} \end{pmatrix} \begin{pmatrix} \omega_1 \\ \vdots \\ \omega_r \end{pmatrix},$$

where the above $r \times r$ matrix is invertible at any $p \in \mathcal{U}$ because the vectors v_{F_i} span $T_p\mathcal{U}$ at every $p \in \mathcal{U}$ and because $y_1, \dots, y_r$ is a system of coordinates on $\mathcal{U}$.

- Since $\mathcal{U}$ is a coordinate neighborhood the closed forms ω_i are exact and we may integrate each of the ω_i to obtain r functions $t_1, \ldots, t_r$, which are defined up to an additive constant. We choose such constants such that p corresponds to $t_1 = \cdots = t_r = 0$. Notice that these functions provide a system of coordinates on $\mathcal{U}$ because $\mathrm{d}t_1 \wedge \cdots \wedge \mathrm{d}t_r \neq 0$ on $\mathcal{U}$. Hence we can, by the "Inverse function Theorem", write the coordinates $(y_1, \ldots, y_r)$ locally, around p, in terms of $(t_1, \ldots, t_r)$.
- By construction, we can write $\partial/\partial t_i = \mathsf{v}_{F_i}$ on $\mathcal{U}$. Therefore the resulting functions $(y_1(t_1, \ldots, t_r), \ldots, y_r(t_1, \ldots, t_r))$ provide the integral curve of v_{F_i} which passes through $\widetilde{p}$ by putting $t_1, \ldots, t_{i-1}, t_{i+1}, \ldots, t_r$ equal to zero. Using the equations $F_i = c_i$ we get the corresponding integral curve of v_{F_i} as a curve in $\mathcal{M}$, via the "Implicit function Theorem". One similarly determines the integral curve that corresponds to a linear combination of the vector fields v_{F_i}.

► In the next example we present the integration of the three-dimensional Euler top in terms of elliptic functions.

Example 3.8 (*The Euler top*)

The three-dimensional Euler top, governed by the system of ODEs

$$\begin{cases} \dot{x}_1 = (a_2 - a_3)\, x_2\, x_3, \\ \dot{x}_2 = (a_3 - a_1)\, x_3\, x_1, \\ \dot{x}_3 = (a_1 - a_2)\, x_1\, x_2, \end{cases}$$

where $a_1, a_2, a_3 \in \mathbb{R}$ is a completely integrable system. We show that it is explicitly solvable in terms of *elliptic functions*.

- We write the system of ODEs of the Euler top in the form

$$\begin{cases} \dot{x}_1 = \alpha_1\, x_2\, x_3, \\ \dot{x}_2 = \alpha_2\, x_3\, x_1, \\ \dot{x}_3 = \alpha_3\, x_1\, x_2, \end{cases} \tag{3.7}$$

  where

$$\alpha_1 := \frac{1}{I_2} - \frac{1}{I_3}, \qquad \alpha_2 := \frac{1}{I_3} - \frac{1}{I_1}, \qquad \alpha_3 := \frac{1}{I_1} - \frac{1}{I_2}.$$

  We assume for definiteness that $I_1 > I_2 > I_3 > 0$, so that $\alpha_1 < 0$, $\alpha_2 > 0$, $\alpha_3 < 0$.
- System (3.7) admits two functionally independent integrals of motion. In particular, the following three functions are integrals of motion:

$$F_1(x) := \alpha_2\, x_3^2 - \alpha_3\, x_2^2, \qquad F_2(x) := \alpha_3\, x_1^2 - \alpha_1\, x_3^2, \qquad F_3(x) := \alpha_1\, x_2^2 - \alpha_2\, x_1^2.$$

  Clearly, only two of them are functionally independent because of $\alpha_1\, F_1 + \alpha_2\, F_2 + \alpha_3\, F_3 = 0$.
- One easily sees that the coordinates x_j satisfy the following decoupled ODEs with coefficients depending on the integrals of motion:

$$\begin{cases} \dot{x}_1^2 = \left(F_3 + \alpha_2\, x_1^2\right)\left(\alpha_3\, x_1^2 - F_2\right), \\ \dot{x}_2^2 = \left(F_1 + \alpha_3\, x_2^2\right)\left(\alpha_1\, x_2^2 - F_3\right), \\ \dot{x}_3^2 = \left(F_2 + \alpha_1\, x_3^2\right)\left(\alpha_2\, x_3^2 - F_1\right). \end{cases}$$

The fact that the polynomials on the right-hand sides of these equations are of degree four implies that the solutions are given by elliptic functions. Note that in the above ODEs one regards F_1, F_2, F_3 as fixed real values.

- Let us introduce the four *Jacobi theta functions*. Let $\tau \in \mathbb{C}$ with $\mathfrak{Im}\ \tau > 0$, $q := \mathrm{e}^{\pi \mathrm{i} \tau}$ and u be a complex variable. The we define:

$$
\begin{aligned}
\vartheta_1(u) &:= 2\sum_{n=0}^{\infty}(-1)^n q^{\left(n+\frac{1}{2}\right)^2}\sin((2n+1)\pi u)\\
&= -\mathrm{i}\sum_{n\in\mathbb{Z}}(-1)^n q^{\left(n+\frac{1}{2}\right)^2}\mathrm{e}^{(2n+1)\pi\mathrm{i}u},\\
\vartheta_2(u) &:= 2\sum_{n=0}^{\infty} q^{\frac{1}{4}(2n+1)^2}\cos((2n+1)\pi u)\\
&= \sum_{n\in\mathbb{Z}} q^{\left(n+\frac{1}{2}\right)^2}\mathrm{e}^{(2n+1)\pi\mathrm{i}u},\\
\vartheta_3(u) &:= 1+2\sum_{n=1}^{\infty} q^{n^2}\cos(2n\pi u)\\
&= \sum_{n\in\mathbb{Z}} q^{n^2}\mathrm{e}^{2n\pi\mathrm{i}u},\\
\vartheta_4(u) &:= 1+2\sum_{n=1}^{\infty}(-1)^n q^{n^2}\cos(2n\pi u)\\
&= \sum_{n\in\mathbb{Z}}(-1)^n q^{n^2}\mathrm{e}^{2n\pi\mathrm{i}u}.
\end{aligned}
$$

We list some important properties of Jacobi theta functions.

(a) Argument $u = 0$ is usually omitted, the corresponding values are called *theta-constants*:

$$\vartheta_2 := \sum_{n\in\mathbb{Z}} q^{(n+\frac{1}{2})^2}, \qquad \vartheta_3 := \sum_{n\in\mathbb{Z}} q^{n^2}, \qquad \vartheta_4 := \sum_{n\in\mathbb{Z}} (-1)^n q^{n^2}.$$

They satisfy the identity $\vartheta_2^2 + \vartheta_4^2 = \vartheta_3^2$.

(b) The four theta functions are quasi double-periodic functions of u and they play a fundamental role in the classical theory of elliptic functions. In particular they are related to the *Jacobi elliptic functions* via the formulas

$$
\begin{aligned}
\mathrm{sn}(u,k) &= \frac{\vartheta_3\,\vartheta_1\,(u/\omega_1)}{\vartheta_2\,\vartheta_4\,(u/\omega_1)},\\
\mathrm{cn}(u,k) &= \frac{\vartheta_4\,\vartheta_2\,(u/\omega_1)}{\vartheta_2\,\vartheta_4\,(u/\omega_1)},\\
\mathrm{dn}(u,k) &= \frac{\vartheta_4\,\vartheta_3\,(u/\omega_1)}{\vartheta_3\,\vartheta_4\,(u/\omega_1)},
\end{aligned}
$$

where $\omega_1 := \pi\,\vartheta_3^2$. The quantity $k^2 := \vartheta_2^4/\vartheta_3^4$ is called *modulus*, while $(k')^2 = 1 - k^2 = \vartheta_4^4/\vartheta_3^4$ is the *complementary modulus*.

(c) The following *relations between squares* hold (there are two linearly independent ones):

$$
\begin{aligned}
\vartheta_1^2(u)\vartheta_4^2 + \vartheta_2^2(u)\vartheta_3^2 - \vartheta_3^2(u)\vartheta_2^2 &= 0, &(3.8)\\
\vartheta_1^2(u)\vartheta_3^2 + \vartheta_2^2(u)\vartheta_4^2 - \vartheta_4^2(u)\vartheta_2^2 &= 0, &(3.9)\\
\vartheta_1^2(u)\vartheta_2^2 + \vartheta_3^2(u)\vartheta_4^2 - \vartheta_4^2(u)\vartheta_3^2 &= 0, &(3.10)\\
\vartheta_2^2(u)\vartheta_2^2 - \vartheta_3^2(u)\vartheta_3^2 + \vartheta_4^2(u)\vartheta_4^2 &= 0. &(3.11)
\end{aligned}
$$

(d) The following *formulas for derivatives* hold:

$$\begin{aligned}
\vartheta_1'(u)\vartheta_2(u) - \vartheta_1(u)\vartheta_2'(u) &= \vartheta_3(u)\vartheta_4(u)\vartheta_2^2,\\
\vartheta_1'(u)\vartheta_3(u) - \vartheta_1(u)\vartheta_3'(u) &= \vartheta_2(u)\vartheta_4(u)\vartheta_3^2,\\
\vartheta_1'(u)\vartheta_4(u) - \vartheta_1(u)\vartheta_4'(u) &= \vartheta_2(u)\vartheta_3(u)\vartheta_4^2,\\
\vartheta_3'(u)\vartheta_2(u) - \vartheta_3(u)\vartheta_2'(u) &= \vartheta_1(u)\vartheta_4(u)\vartheta_4^2,\\
\vartheta_4'(u)\vartheta_2(u) - \vartheta_4(u)\vartheta_2'(u) &= \vartheta_1(u)\vartheta_3(u)\vartheta_3^2,\\
\vartheta_4'(u)\vartheta_3(u) - \vartheta_4(u)\vartheta_3'(u) &= \vartheta_1(u)\vartheta_2(u)\vartheta_2^2.
\end{aligned}$$

In particular, from the above formulas there follows:

$$\frac{\mathrm{d}}{\mathrm{d}u}\frac{\vartheta_1(u)}{\vartheta_4(u)} = \frac{\vartheta_2(u)}{\vartheta_4(u)}\frac{\vartheta_3(u)}{\vartheta_4(u)}\vartheta_4^2, \tag{3.12}$$

$$\frac{\mathrm{d}}{\mathrm{d}u}\frac{\vartheta_2(u)}{\vartheta_4(u)} = -\frac{\vartheta_1(u)}{\vartheta_4(u)}\frac{\vartheta_3(u)}{\vartheta_4(u)}\vartheta_3^2, \tag{3.13}$$

$$\frac{\mathrm{d}}{\mathrm{d}u}\frac{\vartheta_3(u)}{\vartheta_4(u)} = -\frac{\vartheta_1(u)}{\vartheta_4(u)}\frac{\vartheta_2(u)}{\vartheta_4(u)}\vartheta_2^2. \tag{3.14}$$

(e) The following *addition formulas* hold:

$$\begin{aligned}
\vartheta_1(u\pm v)\vartheta_2(u\mp v)\vartheta_3\vartheta_4 &= \vartheta_1(u)\vartheta_2(u)\vartheta_3(v)\vartheta_4(v) \pm \vartheta_3(u)\vartheta_4(u)\vartheta_1(v)\vartheta_2(v),\\
\vartheta_1(u\pm v)\vartheta_3(u\mp v)\vartheta_2\vartheta_4 &= \vartheta_1(u)\vartheta_3(u)\vartheta_2(v)\vartheta_4(v) \pm \vartheta_2(u)\vartheta_4(u)\vartheta_1(v)\vartheta_3(v),\\
\vartheta_1(u\pm v)\vartheta_4(u\mp v)\vartheta_2\vartheta_3 &= \vartheta_1(u)\vartheta_4(u)\vartheta_2(v)\vartheta_3(v) \pm \vartheta_2(u)\vartheta_3(u)\vartheta_1(v)\vartheta_4(v),\\
\vartheta_2(u\pm v)\vartheta_3(u\mp v)\vartheta_2\vartheta_3 &= \vartheta_2(u)\vartheta_3(u)\vartheta_2(v)\vartheta_3(v) \mp \vartheta_1(u)\vartheta_4(u)\vartheta_1(v)\vartheta_4(v),\\
\vartheta_2(u\pm v)\vartheta_4(u\mp v)\vartheta_2\vartheta_4 &= \vartheta_2(u)\vartheta_4(u)\vartheta_2(v)\vartheta_4(v) \mp \vartheta_1(u)\vartheta_3(u)\vartheta_1(v)\vartheta_3(v),\\
\vartheta_3(u\pm v)\vartheta_4(u\mp v)\vartheta_3\vartheta_4 &= \vartheta_3(u)\vartheta_4(u)\vartheta_3(v)\vartheta_4(v) \mp \vartheta_1(u)\vartheta_2(u)\vartheta_1(v)\vartheta_2(v).
\end{aligned}$$

In particular, from the above formulas there follows:

$$\begin{aligned}
&\frac{\vartheta_1(u+v)}{\vartheta_4(u+v)} - \frac{\vartheta_1(u-v)}{\vartheta_4(u-v)} \\
&\quad = \left(\frac{\vartheta_2(u+v)}{\vartheta_4(u+v)}\frac{\vartheta_3(u-v)}{\vartheta_4(u-v)} + \frac{\vartheta_2(u-v)}{\vartheta_4(u-v)}\frac{\vartheta_3(u+v)}{\vartheta_4(u+v)}\right)\frac{\vartheta_1(v)\vartheta_4(v)}{\vartheta_2(v)\vartheta_3(v)},
\end{aligned} \tag{3.15}$$

$$\begin{aligned}
&\frac{\vartheta_2(u+v)}{\vartheta_4(u+v)} - \frac{\vartheta_2(u-v)}{\vartheta_4(u-v)} \\
&\quad = -\left(\frac{\vartheta_1(u+v)}{\vartheta_4(u+v)}\frac{\vartheta_3(u-v)}{\vartheta_4(u-v)} + \frac{\vartheta_1(u-v)}{\vartheta_4(u-v)}\frac{\vartheta_3(u+v)}{\vartheta_4(u+v)}\right)\frac{\vartheta_1(v)\vartheta_3(v)}{\vartheta_2(v)\vartheta_4(v)},
\end{aligned} \tag{3.16}$$

$$\begin{aligned}
&\frac{\vartheta_3(u+v)}{\vartheta_4(u+v)} - \frac{\vartheta_3(u-v)}{\vartheta_4(u-v)} \\
&\quad = -\left(\frac{\vartheta_1(u+v)}{\vartheta_4(u+v)}\frac{\vartheta_2(u-v)}{\vartheta_4(u-v)} + \frac{\vartheta_1(u-v)}{\vartheta_4(u-v)}\frac{\vartheta_2(u+v)}{\vartheta_4(u+v)}\right)\frac{\vartheta_1(v)\vartheta_2(v)}{\vartheta_3(v)\vartheta_4(v)}.
\end{aligned} \tag{3.17}$$

- We can now construct the solution of system (3.7), whose form is indeed analog to formulas (3.12–3.14).
- *First Ansatz* ($F_2 > 0$). We consider the Ansatz

$$x_1 = a\,\frac{\vartheta_2(\nu t)}{\vartheta_4(\nu t)}, \qquad x_2 = b\,\frac{\vartheta_1(\nu t)}{\vartheta_4(\nu t)}, \qquad x_3 = c\,\frac{\vartheta_3(\nu t)}{\vartheta_4(\nu t)}. \tag{3.18}$$

Formulas (3.12–3.14) yield:

$$\dot{x}_1 = -\frac{a\nu}{bc}\,\vartheta_3^2\,x_2\,x_3, \qquad \dot{x}_2 = \frac{b\nu}{ca}\,\vartheta_4^2\,x_3\,x_1, \qquad \dot{x}_3 = -\frac{c\nu}{ab}\,\vartheta_2^2\,x_1\,x_2.$$

Therefore the Ansatz gives a solution, if $a\,b\,c\,\nu > 0$ and

$$\alpha_1 = -\frac{a\,\nu}{b\,c}\,\vartheta_3^2, \qquad \alpha_2 = \frac{b\,\nu}{c\,a}\,\vartheta_4^2, \qquad \alpha_3 = -\frac{c\,\nu}{a\,b}\,\vartheta_2^2.$$

These are three equations for five unknowns (a, b, c, ν, q). We use them to express the three amplitudes a, b, c through the still unknown values ν, q (that is, through ν and theta-constants):

$$a^2 = -\frac{\nu^2\,\vartheta_2^2\,\vartheta_4^2}{\alpha_2\,\alpha_3}, \qquad b^2 = \frac{\nu^2\,\vartheta_2^2\,\vartheta_3^2}{\alpha_1\,\alpha_3}, \qquad c^2 = -\frac{\nu^2\,\vartheta_3^2\,\vartheta_4^2}{\alpha_1\,\alpha_2}. \tag{3.19}$$

The missing two equations come from the integrals. To find them, we use the relations between squares (three of them involving $\vartheta_4^2(x)$, that is, (3.9–3.11); of course, only two of them are independent). Use (3.10), Ansatz (3.18) and then (3.19) to express the amplitudes:

$$1 \;=\; \frac{\vartheta_1^2(\nu\,t)}{\vartheta_4^2(\nu\,t)}\,\frac{\vartheta_2^2}{\vartheta_3^2} + \frac{\vartheta_3^2(\nu\,t)}{\vartheta_4^2(\nu\,t)}\,\frac{\vartheta_4^2}{\vartheta_3^2} \;=\; \frac{x_2^2}{b^2}\,\frac{\vartheta_2^2}{\vartheta_3^2} + \frac{x_3^2}{c^2}\,\frac{\vartheta_4^2}{\vartheta_3^2} = \left(\alpha_3\,x_2^2 - \alpha_2\,x_3^2\right)\frac{\alpha_1}{\nu^2\,\vartheta_3^4}\,.$$

We see that the following quantity is an integral of motion:

$$F_1 = \alpha_2\,x_3^2 - \alpha_3\,x_2^2 = -\frac{\nu^2\,\vartheta_3^4}{\alpha_1}\,. \tag{3.20}$$

There holds $F_1 > 0$. Similarly, starting with (3.9), we find:

$$1 \;=\; \frac{\vartheta_1^2(\nu\,t)}{\vartheta_4^2(\nu\,t)}\,\frac{\vartheta_3^2}{\vartheta_2^2} + \frac{\vartheta_2^2(\nu\,t)}{\vartheta_4^2(\nu\,t)}\,\frac{\vartheta_4^2}{\vartheta_2^2} = \left(\alpha_1\,x_2^2 - \alpha_2\,x_1^2\right)\frac{\alpha_3}{\nu^2\,\vartheta_2^4}\,.$$

Thus, we have found one more integral of motion,

$$F_3 = \alpha_1\,x_2^2 - \alpha_2\,x_1^2 = \frac{\nu^2\,\vartheta_2^4}{\alpha_3}\,, \tag{3.21}$$

with $F_3 < 0$. Finally, we perform a similar computation starting with (3.11):

$$1 \;=\; \frac{\vartheta_3^2(\nu\,t)}{\vartheta_4^2(\nu\,t)}\,\frac{\vartheta_3^2}{\vartheta_4^2} - \frac{\vartheta_2^2(\nu\,t)}{\vartheta_4^2(\nu\,t)}\,\frac{\vartheta_2^2}{\vartheta_4^2} = \left(\alpha_3\,x_1^2 - \alpha_1\,x_3^2\right)\frac{\alpha_2}{\nu^2\,\vartheta_4^4}\,,$$

so that the third integral of motion reads

$$F_2 = \alpha_3\,x_1^2 - \alpha_1\,x_3^2 = \frac{\nu^2\,\vartheta_4^4}{\alpha_2}. \tag{3.22}$$

From this expression we see that our first Ansatz corresponds those the initial data with $F_2 > 0$.

- Integrals (3.20–3.22) are, of course, linearly dependent. Note that $\alpha_1 F_1 + \alpha_2 F_2 + \alpha_3 F_3 = 0$ is equivalent to the identity $\vartheta_2^2 + \vartheta_4^2 = \vartheta_3^2$. These integrals allow us to determine q (or the theta-constants):

$$k^2 := \frac{\vartheta_2^4}{\vartheta_3^4} = -\frac{\alpha_3\,F_3}{\alpha_1\,F_1}, \qquad (k')^2 := 1 - k^2 = \frac{\vartheta_4^4}{\vartheta_3^4} = -\frac{\alpha_2\,F_2}{\alpha_1\,F_1}.$$

Both modules k^2 and $(k')^2$ are positive and therefore lie in $(0, 1)$. For the frequency ν we have the expression $\nu^2\vartheta_3^4 = -\alpha_1\,F_1$. Finally, formulas (3.19–3.22) allow us to express the amplitudes a, b, c in terms of the integrals:

$$a^4 = \frac{F_2\,F_3}{\alpha_2\,\alpha_3}, \qquad b^4 = -\frac{F_1\,F_3}{\alpha_1\,\alpha_3}, \qquad c^4 = -\frac{F_1\,F_2}{\alpha_1\,\alpha_2}.$$

- *Second Ansatz* ($F_2 < 0$). We consider the Ansatz

$$x_1 = a\,\frac{\vartheta_3(\nu\,t)}{\vartheta_4(\nu\,t)}, \qquad x_2 = b\,\frac{\vartheta_1(\nu\,t)}{\vartheta_4(\nu\,t)}, \qquad x_3 = c\,\frac{\vartheta_2(\nu\,t)}{\vartheta_4(\nu\,t)}. \tag{3.23}$$

A computation similar to the one performed for the first Ansatz gives

$$a^4 = -\frac{F_2\,F_3}{\alpha_2\,\alpha_3}, \qquad b^4 = -\frac{F_1\,F_3}{\alpha_1\,\alpha_3}, \qquad c^4 = \frac{F_1\,F_2}{\alpha_1\,\alpha_2}.$$

3.3 The Arnold-Liouville Theorem

► The "Arnold-Liouville Theorem" can be considered as the motivation of Definition 3.1.

- This theorem, in its modern formulation, describes the manifolds, traced out by the flows of completely integrable vector fields, assuming compactness of these manifolds or completeness of these flows on them. It turns out that these manifolds are tori or cylinders and the flow on them is linear.
- The original Liouville's formulation of the theorem holds for symplectic manifolds and it claims that complete integrability implies solvability without mentioning any topological constraint on the manifolds where solution curves are defined. Such theorem can be formulated as follows: *A Hamiltonian system, defined on a symplectic manifold of dimension $2r$, which admits r functionally independent and Poisson involutive integrals of motion is solvable by quadratures.*

► In order to claim and prove the "Arnold-Liouville Theorem" we need two preliminary lemmas and some intermediate notions. Let $(\mathcal{M}, \{\cdot,\cdot\}, \mathtt{F})$ be a completely integrable system of rank $2r$ (recall that $s := n - r$). We denote by $\mathsf{v}_{F_1}, \ldots, \mathsf{v}_{F_s}$ the (completely integrable) Hamiltonian vector fields associated with $\mathtt{F}$. We denote by $\Phi_{t_1}, \ldots, \Phi_{t_s}$ the corresponding (local) Hamiltonian flows. Our considerations are always local and we always assume that $p \in \mathcal{U}_{\mathtt{F}} \cap \mathcal{M}_{(r)}$.

Lemma 3.1

1. *The open subset $\mathcal{U}_{\mathtt{F}} \cap \mathcal{M}_{(r)}$ is invariant under the action of each flow Φ_{t_i}, $i = 1, \ldots, s$.*
2. *The vector fields $\mathsf{v}_{F_1}, \ldots, \mathsf{v}_{F_s}$ define a distribution*

$$\mathcal{D}(p) := \mathsf{span}\{\mathsf{v}_{F_1}(p), \ldots, \mathsf{v}_{F_s}(p)\}, \qquad p \in \mathcal{U}_{\mathtt{F}} \cap \mathcal{M}_{(r)}, \tag{3.24}$$

of rank r that is integrable (in the sense of Frobenius).

Proof. We prove both claims.

1. Let t be any one of the times t_i, $i = 1, \ldots, s$. We have, for small $|t|$, that $\Phi_t^\star F_j = F_j$, $j = 1, \ldots, s$, because v_{F_j} is tangent to the fibers of $\mathtt{F}$. Therefore (see (1.19))

$$\begin{aligned} \Phi_t^\star(\mathsf{d}F_1 \wedge \cdots \wedge \mathsf{d}F_s) &= \mathsf{d}(\Phi_t^\star F_1) \wedge \cdots \wedge \mathsf{d}(\Phi_t^\star F_s) \\ &= \mathsf{d}F_1 \wedge \cdots \wedge \mathsf{d}F_s, \end{aligned}$$

so that the s-form $\mathrm{d}F_1 \wedge \cdots \wedge \mathrm{d}F_s$ is preserved by Φ_t. It follows that $\mathcal{U}_{\mathrm{F}}$ is preserved by Φ_t. But we know that also $\mathcal{M}_{(r)}$ is invariant under the action of these flows. Hence their intersection also, which is the claim.

2. Formula (3.6) implies that the vector fields v_{F_i} define on $\mathcal{U}_{\mathrm{F}} \cap \mathcal{M}_{(r)}$ a distribution of rank r, which is given by (3.24). This distribution is integrable (in the sense of Frobenius) because the vector fields v_{F_i} commute.

The claim is proved. ■

Definition 3.2

The maximal integral manifold of $\mathcal{D}(p)$, $p \in \mathcal{U}_{\mathrm{F}} \cap \mathcal{M}_{(r)}$, is called the **invariant manifold** *of* F *passing through* p. *We denote it by* $\mathcal{F}'_p$.

► Remarks:

- The integrable vector fields v_{F_i} define on $\mathcal{M}$ a generalized distribution, which can also be shown to be integrable, hence we can define $\mathcal{F}'_p$ for any $p \in \mathcal{M}$. However we will only restrict our analysis to $\mathcal{F}'_p$, where $p \in \mathcal{U}_{\mathrm{F}} \cap \mathcal{M}_{(r)}$.
- The invariant manifold of F that passes through p is by definition the immersed submanifold which is traced out by the flow of the integrable vector fields starting at p. In the next lemma we give an alternative description of the invariant manifold $\mathcal{F}'_p$, which on the one hand explains the notation $\mathcal{F}'_p$ (recall that $\mathcal{F}_p$ denotes the fiber of F which passes through p), and on the other hand shows that $\mathcal{F}'_p$ is actually an embedded submanifold of $\mathcal{M}$.

Lemma 3.2

Let $p \in \mathcal{U}_{\mathrm{F}} \cap \mathcal{M}_{(r)}$. The invariant manifold $\mathcal{F}'_p$ of F *is the connected component of $\mathcal{F}_p \cap \mathcal{U}_{\mathrm{F}} \cap \mathcal{M}_{(r)}$, which contains p. In particular, $\mathcal{F}'_p$ is an embedded submanifold of $\mathcal{M}$.*

Proof. Let us denote by $\widetilde{\mathcal{F}}_p$ the connected component of $\mathcal{F}_p \cap \mathcal{U}_{\mathrm{F}} \cap \mathcal{M}_{(r)}$ which contains p. Since $\widetilde{\mathcal{F}}_p$ is a connected submanifold of dimension r, whose tangent space at each of its points p coincides with $\mathcal{D}(p)$, we have that $\widetilde{\mathcal{F}}_p \subseteq \mathcal{F}'_p$. On the other hand $\mathcal{F}'_p \subset \mathcal{U}_{\mathrm{F}} \cap \mathcal{M}_{(r)}$, by definition, while $\mathcal{F}'_p \subseteq \mathcal{F}_p$ since $\mathcal{D}(p)$ is tangent to $\mathcal{F}_p$. Since $\mathcal{F}'_p$ contains p and is connected, it follows that $\mathcal{F}'_p \subseteq \widetilde{\mathcal{F}}_p$. ■

► We now need two technical results which will be used in the proof of next Theorem.

- The following claim holds true: All discrete subgroups of $\mathbb{R}^r$ are isomorphic to $\mathbb{Z}^q$ with $0 \leqslant q \leqslant r$.

- Let $a \in \mathbb{R}^r$ be a fixed vector, fix $\tau \in \mathbb{R}$ and consider the *translational flow*

$$S_\tau : \mathbb{R}^r \longrightarrow \mathbb{R}^r \; : v \mapsto v + \tau\, a.$$

(a) Let $\pi : \mathbb{R}^r \longrightarrow \mathbb{R}^{r-q} \times \mathbb{T}^q$ be the canonical projection. A flow

$$\Psi_\tau : \mathbb{R}^{r-q} \times \mathbb{T}^q \longrightarrow \mathbb{R}^{r-q} \times \mathbb{T}^q$$

is a *translational-type flow* if

$$\Psi_\tau = \pi \circ S_\tau \circ \pi^{-1}.$$

(b) A translational-type flow is *quasi-periodic* if $q = r$: if we denote by $(y_1, \ldots y_r)$ a set of coordinates on $\mathbb{T}^r$ we have:

$$\Psi_\tau : \mathbb{T}^r \longrightarrow \mathbb{T}^r \; : y_i \mapsto y_i + \tau\, a_i \,(\mathrm{mod}\, 1).$$

The following claim holds true: Each orbit of Ψ_τ is dense on $\mathbb{T}^r$ if and only if $k_1\, a_1 + \cdots + k_r\, a_r = 0$, $k_i \in \mathbb{Z}$, implies $k_i = 0$ for all $i = 1, \ldots, r$.

► We are now ready to claim and prove the next statement.

Theorem 3.2 (*Arnold-Liouville*)

Let $p \in \mathcal{U}_\mathrm{F} \cap \mathcal{M}_{(r)}$.

1. *If $\mathcal{F}'_p$ is compact then there exists a diffeomorphism from $\mathcal{F}'_p$ to the torus $\mathbb{T}^r \simeq (\mathbb{R}/\mathbb{Z})^r$.*
2. *If $\mathcal{F}'_p$ is not compact, but Φ_{t_i} is complete on $\mathcal{F}'_p$ for all $i = 1, \ldots, s$, then there exists a diffeomorphism from $\mathcal{F}'_p$ to a cylinder $\mathbb{R}^{r-q} \times \mathbb{T}^q$, with $0 \leqslant q < r$.*

In both cases the diffeomorphism can be chosen in such a way that the vector fields $\mathsf{v}_{F_1}, \ldots, \mathsf{v}_{F_s}$ are mapped to translational-type vector fields. They define quasi-periodic flows in case 1.

Proof. We consider a point $p \in \mathcal{U}_\mathrm{F} \cap \mathcal{M}_{(r)}$. To prove both claims we assume that Φ_{t_i} is complete on $\mathcal{F}'_p$, which is is true in the first case because $\mathcal{F}'_p$ is compact. We also assume that the functions F_i are ordered in such a way that $\mathsf{v}_{F_1}, \ldots, \mathsf{v}_{F_r}$ are the r independent vector fields.

- The completeness and commutativity of $\mathsf{v}_{F_1}, \ldots, \mathsf{v}_{F_r}$ imply that we can define a local action of the Abelian group $\mathbb{R}^r$, with coordinates $t := (t_1, \ldots, t_r) \in \mathbb{R}^r$, on $\mathcal{F}'_p$ by setting

$$\Phi_t : \mathbb{R}^r \times \mathcal{F}'_p \to \mathcal{F}'_p \; : (t, \widetilde{p}) \mapsto (\Phi_{t_1} \circ \cdots \circ \Phi_{t_r})\, (\widetilde{p}). \tag{3.25}$$

(a) The commutativity of the flows Φ_{t_i} implies the group property for the action Φ_t. Indeed, for $\ell := (\ell_1, \ldots, \ell_r) \in \mathbb{R}^r$, we have:

$$\begin{aligned} \Phi_{t+\ell}(\widetilde{p}) &= \left(\Phi_{t_1+\ell_1} \circ \cdots \circ \Phi_{t_r+\ell_r}\right)(\widetilde{p}) \\ &= \left(\Phi_{t_1} \circ \Phi_{\ell_1} \circ \cdots \circ \Phi_{t_r} \circ \Phi_{\ell_r}\right)(\widetilde{p}) \\ &= \left(\Phi_{t_1} \circ \cdots \circ \Phi_{t_r} \circ \Phi_{\ell_1} \circ \cdots \circ \Phi_{\ell_r}\right)(\widetilde{p}) \\ &= \left(\Phi_t \circ \Phi_\ell\right)(\widetilde{p}) \end{aligned}$$

for all $\widetilde{p} \in \mathcal{F}'_p$.

(b) The action Φ_t is transitive. Indeed, for $\widetilde{p} \in \mathcal{F}'_p$ the map $(t, \widetilde{p}) \mapsto \Phi_t(\widetilde{p})$, whose image is a priori the $\mathbb{R}^r$-orbit through $\widetilde{p}$, is in fact surjective. The independence of the vector fields v_{F_i} implies that Φ_t is a local diffeomorphism, and thus its image is an open subset of $\mathcal{F}'_p$. Hence the orbits are disjoint open subsets of $\mathcal{F}'_p$, and so if $\mathcal{F}'_p$ is connected it must consist of a single orbit.

(c) We conclude that $\mathcal{F}'_p$ becomes a homogeneous space $\mathbb{R}^r/\mathbf{G}_{\widetilde{p}}$, where $\mathbf{G}_{\widetilde{p}}$ is the isotropy group of an arbitrary element $\widetilde{p} \in \mathcal{F}'_p$:

$$\mathbf{G}_{\widetilde{p}} := \{t \in \mathbb{R}^r \,:\, \Phi_t(\widetilde{p}) = \widetilde{p}\}.$$

Note that $\dim \mathbb{R}^r = \dim \mathcal{F}'_p = r$, so that $\dim \mathbf{G}_{\widetilde{p}} = 0$. This means that $\mathbf{G}_{\widetilde{p}}$ is a discrete subgroup of $\mathbb{R}^r$. This is consistent with the fact that the action (3.25) is also (locally) free, condition implied by the independence of the vector fields v_{F_i} at any point of $\mathcal{F}'_p$.

- The isotropy group $\mathbf{G}_{\widetilde{p}}$ is generated by $q \leqslant r$ linearly independent vectors over $\mathbb{R}$, say $a_1, \ldots, a_q$:

$$\mathbf{G}_{\widetilde{p}} = \left\{ \sum_{i=1}^{q} m_i\, a_i \,:\, m_i \in \mathbb{Z} \right\} \simeq \mathbb{Z}^q.$$

- For $\widetilde{p} \in \mathcal{F}'_p$ the diffeomorphism $h : \mathbb{R}^r/\mathbf{G}_{\widetilde{p}} \to \mathcal{F}'_p$ is given by

$$h([t_1, \ldots, t_r]) := \Phi_t(\widetilde{p}),$$

where $[t_1, \ldots, t_r] \in \mathbb{R}^r/\mathbf{G}_{\widetilde{p}}$.

- Define the isomorphism $T : \mathbb{R}^r \to \mathbb{R}^r$ by setting $T\,a_i := e_i$, $i = 1, \ldots, r$, where $a_{q+1}, \ldots, a_r$ is a completion to the basis $a_1, \ldots, a_q$ and $e_1, \ldots, e_r$ is the standard basis of $\mathbb{R}^r$.

- The map T maps $\mathbf{G}_{\widetilde{p}}$ isomorphically onto $\{(0, \ldots, 0)\} \times \mathbb{Z}^q \subset \mathbb{R}^r$ and hence it induces a diffeomorphism

$$\widehat{T} : \mathbb{R}^r/\mathbf{G}_{\widetilde{p}} \to \mathbb{R}^r/\{(0, \ldots, 0)\} \times \mathbb{Z}^q \simeq \mathbb{R}^{r-q} \times \mathbb{T}^q.$$

Then $\widehat{T} \circ h^{-1} : \mathcal{F}'_p \to \mathbb{R}^{r-q} \times \mathbb{T}^q$ is a diffeomorphism.

- Now we conclude:
    1. If $\mathcal{F}'_p$ is compact then $q = r$ and $\mathbf{G}_{\widetilde{p}} \simeq \mathbb{Z}^r$, so that $\mathcal{F}'_p$ is diffeomorphic to $\mathbb{R}^r/\mathbb{Z}^r$, that is a torus smoothly embedded into $\mathcal{M}$.
    2. If $\mathcal{F}'_p$ is not compact then $\mathbf{G}_{\widetilde{p}} \simeq \mathbb{Z}^q$ with $q \leqslant r-1$, so that $\mathcal{F}'_p$ is diffeomorphic to $\mathbb{R}^r/\mathbb{Z}^q \simeq \mathbb{R}^{r-q} \times (\mathbb{R}/\mathbb{Z})^q \simeq \mathbb{R}^{r-q} \times \mathbb{T}^q$.
- To conclude the proof we need to show that the restriction to $\mathcal{F}'_p$ of the flow generated by one of the vector fields, say v_{F_1}, is differentiably conjugate to a translational-type flow on $\mathbb{R}^{r-q} \times \mathbb{T}^q$. We denote this flow by φ_τ, $\tau \in \mathbb{R}$.
- For $\widetilde{p} \in \mathcal{F}'_p$ define the flows

$$\Lambda_\tau := h^{-1} \circ \varphi_\tau\big|_{\mathcal{F}'_p} \circ h, \qquad \Psi_\tau := \widehat{T} \circ \Lambda_\tau \circ \widehat{T}^{-1},$$

where explicitly

$$\begin{aligned} \Lambda_\tau([t_1,\ldots,t_r]) &= \left(h^{-1} \circ \varphi_\tau\right) \left(\Phi_{(t_1,t_2,\ldots,t_r)}(\widetilde{p})\right) \\ &= h^{-1} \circ \Phi_{(\tau+t_1,t_2,\ldots,t_r)}(\widetilde{p}) \\ &= [\tau + t_1, t_2, \ldots, t_r]. \end{aligned}$$

- Note that the diffeomorphism $\widehat{T} \circ h^{-1}$ defines the commuting diagram:

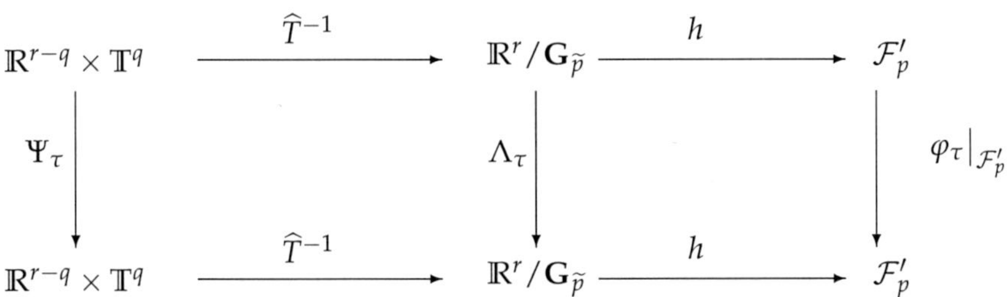

- Consider the flow

$$S_\tau : \mathbb{R}^r \to \mathbb{R}^r : v \mapsto v + \tau\,(T e_1).$$

- We want to prove that Ψ_τ is a translational-type flow defined by S_τ, namely

$$\Psi_\tau = \pi \circ S_\tau \circ \pi^{-1},$$

where $\pi : \mathbb{R}^r \to \mathbb{R}^{r-q} \times \mathbb{T}^q$ is the canonical projection.

- Let $\pi_{\mathbf{G}_{\widetilde{p}}} : \mathbb{R}^r \to \mathbb{R}^r/\mathbf{G}_{\widetilde{p}}$ be the canonical projection and note that by definition of $\widehat{T}$ we have $\pi \circ T = \widehat{T} \circ \pi_{\mathbf{G}_{\widetilde{p}}}$.

- Define $T^{-1}(t_1,\ldots,t_r) := (y_1,\ldots,y_r)$. Then we have:

$$
\begin{aligned}
(\Psi_\tau \circ \pi)(t_1,\ldots,t_r) &= \left(\widehat{T} \circ \Lambda_\tau \circ \widehat{T}^{-1} \circ \pi\right)(t_1,\ldots,t_r) \\
&= \left(\widehat{T} \circ \Lambda_\tau \circ \pi_{\mathbf{G}_{\tilde{p}}} \circ T^{-1}\right)(t_1,\ldots,t_r) \\
&= \left(\widehat{T} \circ \Lambda_\tau \circ \pi_{\mathbf{G}_{\tilde{p}}}\right)(y_1,\ldots,y_r) \\
&= \left(\widehat{T} \circ \Lambda_\tau\right)([t_1,\ldots,t_r]) \\
&= \widehat{T} \circ \pi_{\mathbf{G}_{\tilde{p}}}(\tau + y_1,\ldots,y_r) \\
&= \pi \circ T(\tau\, e_1 + (y_1,\ldots,y_r)) \\
&= \pi(\tau\,(T\, e_1) + (t_1,\ldots,t_r)) \\
&= (\pi \circ S_\tau)(t_1,\ldots,t_r).
\end{aligned}
$$

The Theorem is proved. ■

► Remarks:

- The "Arnold-Liouville Theorem" shows that complete integrability of a Hamiltonian system imposes severe restrictions on the topology of the phase space $\mathcal{M}$. This is the intuitive reason why integrable systems are not generic.
- The real tori that appear in "Arnold-Liouville Theorem" are called *invariant Liouville tori*. "Arnold-Liouville Theorem" assures that the phase space of an integrable Hamiltonian system is foliated into invariant Liouville tori, provided that all trajectories are bounded.
- For most integrable systems that come from classical mechanics (spinning tops, systems of oscillators, penduli, etc.) all fibers $\mathcal{F}_p$ of the momentum map are compact. This does however not mean that each invariant manifold $\mathcal{F}'_p$, with $p \in \mathcal{U}_{\mathrm{F}} \cap \mathcal{M}_{(r)}$, is compact, but it does imply that the flow of the integrable vector fields is complete on each such $\mathcal{F}'_p$. Therefore, each $\mathcal{F}'_p$ is a torus or a cylinder.
- There exist integrable systems which do not satisfy the assumptions of the "Arnold-Liouville Theorem" and the topology of the fibers of the momentum map has to be determined in a different way.

3.3.1 Existence of action-angle coordinates

► The topology of $\mathcal{M}$ resulting from the "Arnold-Liouville Theorem" allows one to introduce some special coordinates, called *action-angle coordinates*, which provide a simple description of the Hamiltonian motion.

► Before providing a formal definition of such variables we describe the construction of an elementary integrable system on the manifold $\mathbb{T}^r \times \mathcal{U}$ where $\mathcal{U}$ is an open subset of $\mathbb{R}^{n-r}$ ($2r \leqslant n$). Such model will be used in the next definition as a local model defined on a neighborhood of a generic compact invariant manifold of the momentum map of a completely integrable system.

- Let $(q_1, \ldots, q_r)$ be a set of local coordinates on $\mathbb{T}^r$.
- Define $\ell := n - 2r$. Let $(p_1, \ldots, p_r, y_1, \ldots, y_\ell)$ be a set of local coordinates on $\mathcal{U}$.
- The set $(q_1, \ldots, q_r, p_1, \ldots, p_r, y_1, \ldots, y_\ell)$ is a set of local coordinates on $\mathbb{T}^r \times \mathcal{U}$.
- We equip $\mathbb{T}^r \times \mathcal{U}$ with a constant Poisson structure $\{\cdot, \cdot\}$ for which all coordinates $y_1, \ldots, y_\ell$ are Casimirs and for which $\{q_i, p_j\} = \delta_{ij}$, with $i, j = 1, \ldots, r$. This Poisson structure can be seen as coming from the canonical Poisson structure of rank $2r$ on $\mathbb{R}^{2r+\ell}$. The corresponding Poisson matrix is

$$\mathcal{X} := \begin{pmatrix} 0_r & 1_r & 0_{r\times\ell} \\ -1_r & 0_r & 0_{r\times\ell} \\ 0_{\ell\times r} & 0_{\ell\times r} & 0_{\ell\times\ell} \end{pmatrix}.$$

- Define $\mathtt{F} := (p_1, \ldots, p_r, y_1, \ldots, y_\ell)$. Then $\mathtt{F}$ is independent (because the set $(p_1, \ldots, p_r, y_1, \ldots, y_\ell)$ is a set of coordinates) and Poisson involutive. Therefore $(\mathbb{T}^r \times \mathcal{U}, \{\cdot, \cdot\}, \mathtt{F})$ is a completely integrable system and each fiber of $\mathtt{F}$ is $\mathbb{T}^r$.
- More generally, define on the above Poisson manifold $(\mathbb{T}^r \times \mathcal{U}, \{\cdot, \cdot\})$ an independent set of functions $\mathbf{G} := (G_1, \ldots, G_{n-r})$ where $G_i \in \mathscr{F}(\mathcal{U})$. Then, by construction, $\mathbf{G}$ is Poisson involutive and therefore $(\mathbb{T}^r \times \mathcal{U}, \{\cdot, \cdot\}, \mathbf{G})$ is a completely integrable system. Each fiber of $\mathbf{G}$ is a disjoint union of tori (indeed, $\mathbf{G}$, viewed as a map $\mathcal{U} \to \mathbb{R}^{n-r}$ needs not be injective). Any integrable vector field v_{G_k}, $k = 1, \ldots, n - r$, is associated with the system of ODEs

$$\begin{cases} \dot{q}_i = \dfrac{\partial G_k}{\partial p_i}, \\ \dot{p}_i = 0, \\ \dot{y}_j = 0, \end{cases} \tag{3.26}$$

  with $i = 1, \ldots, r$ and $j = 1, \ldots, \ell$. The integration of (3.26) is then trivial.

► We now give a formal definition of action-angle coordinates.

Definition 3.3

Let $(\mathcal{M}, \{\cdot, \cdot\}, \mathtt{F})$ be a completely integrable system. Assume that around $p \in \mathcal{M}$

there exists:

1. *An open neighborhood $\mathcal{V}$ of $p \in \mathcal{M}$,*
2. *An open subset $\mathcal{U} \subset \mathbb{R}^{n-r}$,*
3. *A diffeomorphism $\varphi : \mathcal{V} \to \mathbb{T}^r \times \mathcal{U}$,*
4. *A set of functions $\mathtt{G} := (G_1, \dots, G_{n-r})$, $G_i \in \mathscr{F}(\mathcal{U})$, which is independent,*

such that, when $\mathbb{T}^r \times \mathcal{U}$ is equipped with the Poisson structure described above then

(a) *φ is a Poisson map.*

(b) *$\varphi^\star \mathtt{G} = \mathtt{F}$ on $\mathcal{V}$.*

If the above assumptions are satisfied then the canonical coordinates on $\mathcal{V}$ that come from the natural coordinates on $\mathbb{T}^r \times \mathcal{U}$ are called **action-angle coordinates** *and one says that $(\mathcal{M}, \{\cdot,\cdot\}, \mathtt{F})$ admits action-angle coordinates.*

► Remarks:

- In the formula $\varphi^\star \mathtt{G} = \mathtt{F}$ we view the elements of $\mathtt{G}$ as functions on $\mathbb{T}^r \times \mathcal{U}$, so to be more precise, this formula should be written as $\varphi^\star \pi^\star \mathtt{G} = \mathtt{F}$, where $\pi : \mathbb{T}^r \times \mathcal{U} \to \mathcal{U}$ is the projection on the second component.
- One can say that after a Poisson change of coordinates the neighborhood $\mathcal{V}$ looks like the Poisson manifold $\mathbb{T}^r \times \mathcal{U}$ described in the construction above.

► The following theorem gives condition under which a completely integrable system admits action-angle coordinates.

Theorem 3.3 (*Arnold*)

Let $p \in \mathcal{U}_{\mathtt{F}} \cap \mathcal{M}_{(r)}$ and assume that $\mathcal{F}'_p$ is compact. Then $(\mathcal{M}, \{\cdot,\cdot\}, \mathtt{F})$ admits action-angle coordinates around p.

No Proof.

► In general it is difficult to compute explicit action-angle variables for a given integrable system. It is already non-trivial to find whether or not some of the fibers of the momentum map of the integrable system are compact and to localize them. Furthermore the fibration by Liouville tori is in general not trivial and global action-angle variables do not exist.

Example 3.9 (*n-dimensional isotropic oscillator*)

Consider the isotropic oscillator of Example 3.1 with all ν_i equal to 1. In this case the Poisson manifold is the canonical vector space $\mathbb{R}^{2n}$ which has maximal rank $2r = 2n$.

- We have that $\mathtt{F} := (F_1, \ldots, F_r)$, with
$$F_i := \frac{1}{2}\left(p_i^2 + q_i^2\right), \qquad i = 1, \ldots, r,$$
is Poisson involutive and independent.
- In the notation of Definition 3.3 we take $\mathcal{U} := (\mathbb{R}_{>0})^r$ and $\mathcal{V} := (\mathbb{R}^2 \setminus \{(0,0)\})^r$ and we take for $\mathtt{G} := (G_1, \ldots, G_r)$ the standard coordinates on $(\mathbb{R}_{>0})^r$.
- We define a map $\varphi^{-1} : \mathbb{T}^r \times \mathcal{U} \to \mathcal{V}$ by setting
$$(\theta_1, \ldots, \theta_r, G_1, \ldots, G_r) \mapsto \sqrt{2\,G_i}\,(\cos\theta_i, \sin\theta_i)_{i=1,\ldots,r}\,.$$
It is clear that φ^{-1} and its inverse φ are diffeomorphisms.
- We need to verify that φ is a Poisson map and that $\varphi^\star \mathtt{G} = \mathtt{F}$ on $\mathcal{V}$. In terms of φ^{-1} this means that we need to verify that φ^{-1} is a Poisson map and that $(\varphi^{-1})^\star\,\mathtt{F} = \mathtt{G}$ on $\mathbb{T}^r \times \mathcal{U}$.
- Note that
$$\left(\varphi^{-1}\right)^\star F_i = G_i\left(\cos^2\theta_i + \sin^2\theta_i\right) = G_i, \qquad i = 1, \ldots, r,$$
thus φ^{-1} preserves the integrals of motion.
- Since $\{G_i, \theta_j\} = \delta_{ij}$, while the other brackets vanish we have
$$\begin{aligned} \left\{\left(\varphi^{-1}\right)^\star q_i, \left(\varphi^{-1}\right)^\star p_j\right\} &= \left\{\sqrt{2\,G_i}\cos\theta_i, \sqrt{2\,G_j}\sin\theta_j\right\} \\ &= \left(\frac{\sqrt{2\,G_j}}{\sqrt{2\,G_i}}\cos\theta_i\cos\theta_j + \frac{\sqrt{2\,G_i}}{\sqrt{2\,G_j}}\sin\theta_i\sin\theta_j\right)\delta_{ij} \\ &= \delta_{ij} = \left(\varphi^{-1}\right)^\star \{q_i, p_j\}, \end{aligned}$$
for $i, j = 1, \ldots, r$. Similarly,
$$\left\{\left(\varphi^{-1}\right)^\star q_i, \left(\varphi^{-1}\right)^\star q_j\right\} = 0 = \left(\varphi^{-1}\right)^\star \{q_i, q_j\},$$
and
$$\left\{\left(\varphi^{-1}\right)^\star p_i, \left(\varphi^{-1}\right)^\star q_j\right\} = 0 = \left(\varphi^{-1}\right)^\star \{p_i, p_j\}.$$
We conclude that φ^{-1} and φ are Poisson morphisms.

4

R-brackets and r-brackets

4.1 Introduction

► The theory of integrable systems gave birth to interesting constructions, which yield new Lie-Poisson structures on a given finite-dimensional Lie algebra $\mathfrak{g}$, upon using some extra structure on $\mathfrak{g}$ (roughly speaking a matrix). These ideas date back to the world renowned "St. Petersburg School" represented by Faddeev, Reyman, Sklyanin, Semenov-Tian-Shansky and others.

► There are two distinct formalisms, depending on whether this matrix is viewed as a linear map $R : \mathfrak{g} \to \mathfrak{g}$, in which case R is called an *R-matrix*, or as an element $r \in \mathfrak{g} \otimes \mathfrak{g}$, in which case r is called an *r-matrix*. More precisely:

1. For a given linear map $R : \mathfrak{g} \to \mathfrak{g}$, which satisfies a proper condition, a new Lie bracket on $\mathfrak{g}$ is defined by letting
$$[\mathsf{X}, \mathsf{Y}]_R := \frac{1}{2}([R\mathsf{X}, \mathsf{Y}] + [\mathsf{X}, R\mathsf{Y}])$$
for all $\mathsf{X}, \mathsf{Y} \in \mathfrak{g}$. The new Lie bracket on $\mathfrak{g}$ leads to a Poisson structure on $\mathfrak{g}^*$, which is referred to as an *R-bracket* on $\mathfrak{g}^*$. Upon identifying $\mathfrak{g}$ with $\mathfrak{g}^*$, the R-bracket on $\mathfrak{g}^*$ becomes a Lie-Poisson structure on $\mathfrak{g}$.
2. A tensor $r \in \mathfrak{g} \otimes \mathfrak{g}$ can be viewed as a zero-chain on $\mathfrak{g}$, with values in $\mathfrak{g} \otimes \mathfrak{g}$, so its coboundary admits as transpose a linear map $\gamma^\top : \mathfrak{g}^* \otimes \mathfrak{g}^* \to \mathfrak{g}^*$. When r satisfies a condition, which will be explained in detail, $\gamma^\top$ defines a Lie algebra structure on $\mathfrak{g}^*$, hence leads directly to a Lie-Poisson bracket on $\mathfrak{g}$ (i.e., no identification of $\mathfrak{g}$ with $\mathfrak{g}^*$ is needed to define this Lie-Poisson structure on $\mathfrak{g}$). This Lie-Poisson bracket is called an *r-bracket*.

One can show that R-brackets and r-brackets are related.

4.2 R-brackets

► In Chapter 2 we have seen that a Hamiltonian vector field that can be written in the Lax form
$$\dot{\mathsf{X}} = [\mathsf{Y}, \mathsf{X}], \tag{4.1}$$
where X and Y are linear operators which depend on the dynamical variables, admits each coefficient of the characteristic polynomial of X as an integral of motion. As a matter of fact almost all (perhaps, all) known completely integrable systems possess a Lax representation of equations of motion.

► Let $\mathfrak{g} \subseteq \mathfrak{gl}(n,\mathbb{R})$ (resp. $\mathfrak{g} \subseteq \mathfrak{gl}(n,\mathbb{R})\left[\lambda,\lambda^{-1}\right]$) and $X \in \mathfrak{g}$, $Y : \mathfrak{g} \to \mathfrak{gl}(n,\mathbb{R})$ (resp. $\mathfrak{g} \subseteq \mathfrak{gl}(n,\mathbb{R})\left[\lambda,\lambda^{-1}\right]$). Suppose that there exists a Poisson structure on $\mathfrak{g}$ such that a Hamiltonian vector field takes the form (4.1). Then we say that (4.1) is a *Lax equation* (resp. *Lax equation with spectral parameter*).

- Let $\mathfrak{g} = \mathfrak{gl}(n,\mathbb{R})\left[\lambda,\lambda^{-1}\right]$. Then a Lax equation with spectral parameter takes the form
$$\dot{\mathsf{X}}(\lambda) = [\,\mathsf{Y}(\lambda),\mathsf{X}(\lambda)\,], \tag{4.2}$$
where
$$\mathsf{X}(\lambda) = \sum_i \lambda^i \mathsf{X}_i, \qquad \mathsf{X}_i \in \mathfrak{gl}(n,\mathbb{R}),$$
and
$$\mathsf{Y}(\lambda) = \sum_i \lambda^i \mathsf{Y}_i,$$
where each Y_i is a polynomial function of the entries of some (or all) of the X_i.

- A proof similar to the proof of Theorem 2.12 shows that all coefficients of the characteristic polynomial of $\mathsf{X}(\lambda)$, called *spectral invariants*, are integrals of motion of (4.2).

Example 4.1 (*Lax equation with spectral parameter for the Euler top*)

We know from Example 2.16 that the system of ODEs of the Euler top can be written as a Lax equation on $\mathfrak{so}(3)$. One can show that there exists a more effective Lax equation on the loop algebra $\mathfrak{so}(3)[\lambda,\lambda^{-1}]$.

- In Example 2.16 we defined the matrices $\mathsf{X},\Omega \in \mathfrak{so}(3)$ with entries respectively given by
$$\mathsf{X}_{ij} := \varepsilon_{ijk}\, x_k, \qquad \Omega_{ij} := \varepsilon_{ijk}\,\omega_k = \varepsilon_{ijk}\,\frac{x_k}{I_k}.$$
and we found that the equations governing the Euler top are equivalent to the matrix equation
$$\dot{\mathsf{X}} = [\,\mathsf{X},\Omega\,]. \tag{4.3}$$

- Introduce a diagonal matrix $J := \mathsf{diag}(J_1, J_2, J_3)$ where
$$J_k := \frac{1}{2}(I_i + I_j - I_k),$$
where (ijk) is any cyclic permutation of (123). Then we have:
$$\mathsf{X} = J\,\Omega + \Omega\,J.$$

- Define
$$\mathsf{X}(\lambda) := J^2 + \frac{\mathsf{X}}{\lambda}, \qquad \mathsf{Y}(\lambda) := \lambda\,J + \Omega,$$
where $\lambda \in \mathbb{C}$ is an arbitrary parameter. It is easy to check that the Lax equation
$$\dot{\mathsf{X}}(\lambda) = [\,\mathsf{X}(\lambda),\mathsf{Y}(\lambda)\,] \tag{4.4}$$
is equivalent to the equations of motion of the Euler top. We have
$$\dot{\mathsf{X}}(\lambda) = \frac{\dot{\mathsf{X}}}{\lambda},$$

and

$$[\mathsf{X}(\lambda), \mathsf{Y}(\lambda)] = [J^2, \Omega] + [\mathsf{X}, J] + \frac{1}{\lambda}[\mathsf{X}, \Omega].$$

But $[J^2, \Omega] = -[\mathsf{X}, J]$, so that (4.4) reduces to (4.3).

- We observe that the spectral invariants of $\mathsf{X}(\lambda)$ provide all integrals of motion. In particular,

$$\text{Trace}\,(\mathsf{X}(\lambda))^2 = \text{Trace}\, J^4 - \frac{(B(x))^2}{\lambda^2},$$

and

$$\text{Trace}\,(\mathsf{X}(\lambda))^3 = \text{Trace}\, J^6 - \frac{3}{\lambda^2}\left(\frac{1}{4}\,(B(x))^2\text{Trace}\, J^2 - I_1\, I_2\, I_3\, H(x)\right),$$

where

$$H(x) := \frac{1}{2}\left(\frac{x_1^2}{I_1} + \frac{x_2^2}{I_2} + \frac{x_3^2}{I_3}\right), \qquad B(x) := \frac{1}{2}\left(x_1^2 + x_2^2 + x_3^2\right).$$

► The existence of many integrals of motion for a Lax equation indicates that Lax equations should play a special role in the theory of completely integrable systems. Our tasks are:

- To show how to construct Lax equations whose spectral invariants are in Poisson involution w.r.t. some new Lie-Poisson structure.
- To show that - under some assumptions - even if these spectral invariants may not necessarily form a set that is large enough to make the system completely integrable, the Lax equation can be integrated explicitly upon using the underlying Lie group.

► In what follows we consider $\mathfrak{g}$ as a quadratic Lie algebra. More precisely, $\mathfrak{g}$ is a n-dimensional real (or complex) Lie algebra, with Lie bracket $[\cdot,\cdot]$, equipped with a non-degenerate pairing $\langle \cdot \,|\, \cdot \rangle$, which is Ad-invariant (we can take $\langle \cdot \,|\, \cdot \rangle$ as the Killing form of $\mathfrak{g}$ if $\mathfrak{g}$ is semi-simple or simple). The space $\mathfrak{g}^*$ is the dual of $\mathfrak{g}$. Note that $\mathfrak{g}$ can be also a loop algebra.

► In Chapter 2 (see Theorem 2.10) we have shown the following result. Let $\mathsf{X} \in \mathfrak{g}$ and $H \in \mathscr{F}(\mathfrak{g})$. On the Lie-Poisson manifold $(\mathfrak{g}, \{\cdot,\cdot\}_{\text{LP}})$ the Hamiltonian vector field v_H at X takes the form

$$\mathsf{v}_H(\mathsf{X}) = [\nabla H(\mathsf{X}), \mathsf{X}]. \tag{4.5}$$

- It is tempting to compare (4.1) with the Hamiltonian equations of motion (4.5) associated with the Lie-Poisson bracket, but the drawback of the Lax equation (4.5) is that all spectral invariants of X are Casimirs of $(\mathfrak{g}, \{\cdot,\cdot\}_{\text{LP}})$.
- We will see that if instead of using the Lie-Poisson bracket on $\mathfrak{g}$ we properly define a new bracket, then we can arrive at some interesting conclusions about the real nature of complete integrability of vector fields (4.5).

- To implement this idea we shall use the notion of the so called *R-matrix*, which allows one to define another structure on $\mathfrak{g}$ (and on $\mathfrak{g}^*$ by duality). The interplay of these two different structures on the same space proves to be a key property of Lax equations.

4.2.1 Double Lie algebras and linear R-brackets

▶ We start with the following definition.

Definition 4.1

*Let $R \in \mathrm{End}(\mathfrak{g})$ be a linear operator. Then R defines an R-**bracket** on $\mathfrak{g}$, $[\cdot,\cdot]_R : \mathfrak{g} \times \mathfrak{g} \to \mathfrak{g}$, if*

$$[X,Y]_R := \frac{1}{2}([RX,Y]+[X,RY]) \tag{4.6}$$

*is a Lie bracket on $\mathfrak{g}$ for all $X,Y \in \mathfrak{g}$. The resulting Lie algebra $\mathfrak{g}$, equipped with $[\cdot,\cdot]$ and $[\cdot,\cdot]_R$, $(\mathfrak{g},[\cdot,\cdot],[\cdot,\cdot]_R)$, is called **double Lie algebra** and R is called **R-matrix**.*

▶ Let $Q_R : \mathfrak{g} \times \mathfrak{g} \to \mathfrak{g}$ be the bilinear map defined by

$$Q_R(X,Y) := [RX,RY] - R([RX,Y]+[X,RY]), \qquad X,Y \in \mathfrak{g}. \tag{4.7}$$

Then the following statement holds.

Theorem 4.1

$[\cdot,\cdot]_R$ defines a Lie bracket if and only if

$$[Q_R(X,Y),Z] + \circlearrowleft (X,Y,Z) = 0, \qquad \forall X,Y,Z \in \mathfrak{g}. \tag{4.8}$$

Proof. Let us show that $[\cdot,\cdot]_R$ satisfies the Jacobi identity

$$[X,[Y,Z]_R]_R + \circlearrowleft (X,Y,Z) = 0, \qquad \forall X,Y,Z \in \mathfrak{g}, \tag{4.9}$$

if and only if condition (4.8) holds true. Using (4.6) the Jacobi identity (4.9) is

$$\begin{aligned}&[RX,[Y,Z]_R]+[RY,[Z,X]_R]+[RZ,[X,Y]_R]\\ &\quad +[X,R[Y,Z]_R]+[Y,R[Z,X]_R]+[Z,R[X,Y]_R] + \circlearrowleft (X,Y,Z) = 0. \end{aligned} \tag{4.10}$$

Expanding the terms in the first line and using the Jacobi identity for $[\cdot,\cdot]$ we can write

$$\begin{aligned}&[RX,[Y,Z]_R]+[RY,[Z,X]_R]+[RZ,[X,Y]_R]\\ &\quad = -\frac{1}{2}([X,[RY,RZ]]+[Y,[RZ,RX]]+[Z,[RX,RY]]).\end{aligned}$$

They combine with the terms in the second line of (4.10) to give

$$\left[X, R[Y,Z]_R - \frac{1}{2}[RY, RZ] \right] + \circlearrowleft (X,Y,Z) = 0,$$

which is

$$[X, R([RY,Z] + [Y,RZ]) - [RY,RZ]] + \circlearrowleft (X,Y,Z) = 0,$$

that is (4.8). ■

► Remarks:

- The necessary and sufficient condition (4.8) is usually replaced by sufficient conditions, which are bilinear instead of trilinear. The simplest sufficient condition is the so called *(classical) Yang-Baxter equation*

$$Q_R(X,Y) = 0 \qquad \forall X, Y \in \mathfrak{g}, \tag{4.11}$$

  which yields a particular solution to (4.8).

- Another important sufficient condition is the *modified Yang-Baxter equation*

$$Q_R(X,Y) = -c\,[X,Y] \qquad \forall X, Y \in \mathfrak{g}, \tag{4.12}$$

  where c is a constant which can always be rescaled to ± 1 if $\mathfrak{g}$ is a real Lie algebra.

► Recall that a Lie algebra $\mathfrak{g}$ equipped with the Lie bracket $[\cdot,\cdot]$ induces a Lie-Poisson structure on $\mathfrak{g}^*$ whose bracket is

$$\{F,G\}_{\mathrm{LP}}(X^*) := \langle X^*, [\,dF(X^*), dG(X^*)\,] \rangle, \qquad F, G \in \mathscr{F}(\mathfrak{g}^*),$$

where $dF(X^*)$ and $dG(X^*)$ are interpreted as elements of $\mathfrak{g}$ when computing the bracket. We denoted by $(\mathfrak{g}^*, \{\cdot,\cdot\}_{\mathrm{LP}})$ the corresponding Lie-Poisson manifold. We now define a new Lie-Poisson structure on $\mathfrak{g}^*$ induced by the R-bracket.

Definition 4.2

Let $[\cdot,\cdot]_R$ be an R-bracket on $\mathfrak{g}$. For any $F, G \in \mathscr{F}(\mathfrak{g}^*)$ the Lie-Poisson bracket induced by $[\cdot,\cdot]_R$,

$$\begin{aligned} \{F,G\}_R(X^*) &:= \frac{1}{2}\langle X^*, [\,R(dF(X^*)), dG(X^*)\,] \rangle \\ &+ \frac{1}{2}\langle X^*, [\,dF(X^*), R(dG(X^*))\,] \rangle, \end{aligned} \tag{4.13}$$

is called **linear R-bracket** *on $\mathfrak{g}^*$. We denote by $(\mathfrak{g}^*, \{\cdot,\cdot\}_R)$ the corresponding*

Lie-Poisson manifold.

4.2.2 Lie algebra splitting

▶ We now define an important class of double Lie algebras. We start with the following definition.

Definition 4.3

We say that $\mathfrak{g}$ *admits a* **Lie algebra splitting** *if* $\mathfrak{g}$ *is a direct sum (as a vector space) of two Lie subalgebras* $\mathfrak{g}_+, \mathfrak{g}_-$*:*

$$\mathfrak{g} = \mathfrak{g}_+ \oplus \mathfrak{g}_-. \tag{4.14}$$

We introduce the projection operators by $\pi_\pm : \mathfrak{g} \to \mathfrak{g}_\pm$*:*

$$\mathsf{X}_\pm := \pi_\pm(\mathsf{X}), \qquad \mathsf{X} \in \mathfrak{g}.$$

▶ Remarks:

- The splitting (4.14) leads to a direct sum decomposition of $\mathfrak{g}^*$:

  $$\mathfrak{g}^* = \operatorname{Ann}\mathfrak{g}_+ \oplus \operatorname{Ann}\mathfrak{g}_-,$$

  where

  $$\operatorname{Ann}\mathfrak{g}_\pm := \{\mathsf{X}^* \in \mathfrak{g}^* \, : \, \langle \mathsf{X}^*, \mathsf{X} \rangle = 0 \, \forall \mathsf{X} \in \mathfrak{g}_\pm\}$$

  is called *annihilator* of $\mathfrak{g}_\pm$. Therefore $\operatorname{Ann}\mathfrak{g}_\pm = \mathfrak{g}^*_\mp$.

- Dualizing the projection maps $\pi_\pm$ we find two injective linear maps $\pi^*_\pm : \mathfrak{g}^*_\pm \to \mathfrak{g}^*$, which are explicitly given by

  $$\langle \pi^*_\pm(\mathsf{X}^*), \mathsf{X} \rangle := \langle \mathsf{X}^*, \mathsf{X}_\pm \rangle, \qquad \mathsf{X}^* \in \mathfrak{g}^*_\pm, \mathsf{X} \in \mathfrak{g}.$$

  Thus $\pi^*_\pm(\mathsf{X}^*)$ are the natural extensions of $\mathsf{X}^* \in \mathfrak{g}^*_\pm$ to a linear function on $\mathfrak{g}$.

Theorem 4.2

Let $R \in \operatorname{End}(\mathfrak{g})$ *be defined by*

$$R := \pi_+ - \pi_-. \tag{4.15}$$

Then $[\,\cdot,\cdot\,]_R$ *is an R-bracket, so that* $(\mathfrak{g}, [\,\cdot,\cdot\,], [\,\cdot,\cdot\,]_R)$ *is a double Lie algebra. In particular,*

$$[\mathsf{X},\mathsf{Y}]_R = [\mathsf{X}_+,\mathsf{Y}_+] - [\mathsf{X}_-,\mathsf{Y}_-], \qquad \mathsf{X},\mathsf{Y} \in \mathfrak{g}.$$

Proof. We immediately find that

$$\begin{aligned} [X,Y]_R &= \frac{1}{2}([X_+ - X_-, Y_+ + Y_-] + [X_+ + X_-, Y_+ - Y_-]) \\ &= [X_+, Y_+] - [X_-, Y_-]. \end{aligned}$$

Then we show that $R := \pi_+ - \pi_-$ satisfies the modified Yang-Baxter equation. We have

$$R[X,Y]_R = [X_+, Y_+] + [X_-, Y_-],$$

and

$$[RX, RY] = [X_+, Y_+] - [X_+, Y_-] - [X_-, Y_+] + [X_-, Y_-],$$

which combine into

$$Q_R(X,Y) = [RX, RY] - R([RX, Y] + [X, RY]) = -[X,Y],$$

which is the modified Yang-Baxter equation (4.12) with $c = 1$. ■

▶ We can now consider the Lie-Poisson structure on $\mathfrak{g}^*$, when $\mathfrak{g} = \mathfrak{g}_+ \oplus \mathfrak{g}_-$, induced by the R-bracket (4.15).

- If we consider formula (4.13) with $R := \pi_+ - \pi_-$ we find that for any $F, G \in \mathscr{F}(\mathfrak{g}^*)$ the Lie-Poisson R-bracket on $\mathfrak{g}^*$ is

$$\begin{aligned} \{F,G\}_R(X^*) &:= \langle X^*, [(dF(X^*))_+, (dG(X^*))_+] \rangle \\ &- \langle X^*, [(dF(X^*))_-, (dG(X^*))_-] \rangle. \end{aligned}$$

- Furthermore, one can prove the following. Denote by $\{\cdot,\cdot\}_\pm$ the Lie-Poisson structures on $\mathfrak{g}^*_\pm$. Then the map

$$\pi_+^* : (\mathfrak{g}_+^*, \{\cdot,\cdot\}_+) \to (\mathfrak{g}^*, \{\cdot,\cdot\}_R)$$

is a Poisson morphism, while the map

$$\pi_-^* : (\mathfrak{g}_-^*, \{\cdot,\cdot\}_-) \to (\mathfrak{g}^*, \{\cdot,\cdot\}_R),$$

is an anti-Poisson morphism.

Example 4.2

Let $\mathfrak{g}$ be a finite-dimensional Lie algebra. Then the corresponding loop algebra is

$$\mathfrak{g}\left[\lambda, \lambda^{-1}\right] := \left\{X(\lambda) = \sum_i \lambda^i X_i \, : \, X_i \in \mathfrak{g}, \, \lambda \in \mathbb{C}\right\} = \bigoplus_i \lambda^i \mathfrak{g}.$$

- Then $\mathfrak{g}\left[\lambda,\lambda^{-1}\right]$ admits the natural splitting

$$\mathfrak{g}\left[\lambda,\lambda^{-1}\right]=\mathfrak{g}_{+}\left[\lambda\right]\oplus\mathfrak{g}_{-}\left[\lambda^{-1}\right],$$

where

$$\mathfrak{g}_{+}\left[\lambda\right]:=\bigoplus_{i\geqslant 0}\lambda^{i}\mathfrak{g},\qquad \mathfrak{g}_{-}\left[\lambda^{-1}\right]:=\bigoplus_{i<0}\lambda^{i}\mathfrak{g}.$$

- For $\mathsf{X}(\lambda)\in\mathfrak{g}\left[\lambda,\lambda^{-1}\right]$ we can define

$$\mathsf{X}_{+}(\lambda):=\pi_{+}(\mathsf{X}(\lambda))=\sum_{i\geqslant 0}\lambda^{i}\mathsf{X}_{i}\in\mathfrak{g}_{+}\left[\lambda\right],$$
$$\mathsf{X}_{-}(\lambda):=\pi_{-}(\mathsf{X}(\lambda))=\sum_{i<0}\lambda^{i}\mathsf{X}_{i}\in\mathfrak{g}_{-}\left[\lambda^{-1}\right].$$

4.2.3 *Poisson involutivity and factorization theorems*

► Let **G** be a connected matrix Lie group with Lie algebra $\mathfrak{g}$. Note that the assumption that **G** is a matrix Lie group is not necessary but it simplifies computations. It us useful to refresh some formulas we introduced in Chapter 1.

- The adjoint action of **G** on $\mathfrak{g}$ is the map

$$\mathsf{Ad}:\mathbf{G}\to\mathbf{GL}(\mathfrak{g}):h\mapsto\mathsf{Ad}_{h},$$

where $\mathsf{Ad}_{h}\mathsf{X}=h\mathsf{X}h^{-1}$, $h\in\mathbf{G},\mathsf{X}\in\mathfrak{g}$. The coadjoint action of **G** on $\mathfrak{g}^{*}$,

$$\mathsf{Ad}^{*}:\mathbf{G}\to\mathbf{GL}(\mathfrak{g}^{*}):h\mapsto\mathsf{Ad}_{h}^{*},$$

is defined by

$$\langle\,\mathsf{Ad}_{h}^{*}\mathsf{X}^{*},\mathsf{X}\,\rangle=\langle\,\mathsf{X}^{*},\mathsf{Ad}_{h^{-1}}\mathsf{X}\,\rangle,\qquad \mathsf{X}^{*}\in\mathfrak{g}^{*},\mathsf{X}\in\mathfrak{g}.$$

- The adjoint action of $\mathfrak{g}$ on $\mathfrak{g}$ is the map

$$\mathsf{ad}_{\mathsf{X}}:\mathfrak{g}\to\mathfrak{g}:\mathsf{X}\mapsto\mathsf{ad}_{\mathsf{X}}=[\mathsf{X},\cdot\,],$$

while the coadjoint action of $\mathfrak{g}$ on $\mathfrak{g}^{*}$,

$$\mathsf{ad}_{\mathsf{X}}^{*}:\mathfrak{g}\to\mathfrak{g}^{*}:\mathsf{X}\mapsto\mathsf{ad}_{\mathsf{X}}^{*},$$

is such that

$$\langle\,\mathsf{ad}_{\mathsf{X}}^{*}\mathsf{X}^{*},\mathsf{Y}\,\rangle=-\langle\,\mathsf{X}^{*},\mathsf{ad}_{\mathsf{X}}\mathsf{Y}\,\rangle=\langle\,\mathsf{X}^{*},[\,\mathsf{Y},\mathsf{X}\,]\,\rangle,\qquad \mathsf{X}^{*}\in\mathfrak{g}^{*},\mathsf{X},\mathsf{Y}\in\mathfrak{g}.\quad(4.16)$$

- A function $H\in\mathscr{F}(\mathfrak{g}^{*})$ belongs to the algebra $\mathscr{F}_{\mathbf{G}}(\mathfrak{g}^{*})$ of Ad^{*}-invariant functions if

$$H\left(\mathsf{Ad}_{h}^{*}\mathsf{X}^{*}\right)=H(\mathsf{X}^{*})\qquad\forall h\in\mathbf{G},\mathsf{X}^{*}\in\mathfrak{g}^{*}.$$

Let $H\in\mathscr{F}_{\mathbf{G}}(\mathfrak{g}^{*})$. Then (see Theorem 1.1):

(a) For any $\mathsf{X}^* \in \mathfrak{g}^*$ and for any $\mathsf{X} \in \mathfrak{g}$ we have

$$\langle \mathsf{X}^*, [\,\mathrm{d}H(\mathsf{X}^*), \mathsf{X}\,] \rangle = 0, \tag{4.17}$$

i.e., for any $\mathsf{X}^* \in \mathfrak{g}^*$ we have

$$\mathrm{ad}^*_{\mathrm{d}H(\mathsf{X}^*)} \mathsf{X}^* = 0.$$

(b) For any $\mathsf{X}^* \in \mathfrak{g}^*$ and for any $h \in \mathbf{G}$ we have

$$\mathrm{d}H(\mathrm{Ad}^*_h \mathsf{X}^*) = \mathrm{Ad}_h(\mathrm{d}H(\mathsf{X}^*)). \tag{4.18}$$

► We now claim the Semenov-Tian-Shansky Theorem on $\mathfrak{g}^*$, which is indeed the motivation of the definition of a double Lie algebra.

Theorem 4.3 (*Semenov-Tian-Shansky*)

Let $(\mathfrak{g}, [\,\cdot,\cdot\,], [\,\cdot,\cdot\,]_R)$ be a double Lie algebra. Then:

1. *Any two functions $F, H \in \mathscr{F}_{\mathbf{G}}(\mathfrak{g}^*)$ are in involution w.r.t. the Lie-Poisson R-bracket (4.13), i.e., $\{F, H\}_R = 0$.*
2. *Let $H \in \mathscr{F}_{\mathbf{G}}(\mathfrak{g}^*)$. The Hamiltonian vector field $\mathsf{v}_H := \{\,\cdot, H\}_R$ takes the form*
$$\mathsf{v}_H(\mathsf{X}^*) = \frac{1}{2}\,\mathrm{ad}^*_{\mathsf{B}} \mathsf{X}^*, \qquad \mathsf{B} := R(\mathrm{d}H(\mathsf{X}^*)). \tag{4.19}$$
In particular, if $\mathfrak{g} = \mathfrak{g}_+ \oplus \mathfrak{g}_-$ and $R := \pi_+ - \pi_-$, one has
$$\mathsf{v}_H(\mathsf{X}^*) = \pm\,\mathrm{ad}^*_{\mathsf{B}_\pm} \mathsf{X}^*, \qquad \mathsf{B}_\pm := (\mathrm{d}H(\mathsf{X}^*))_\pm. \tag{4.20}$$

Proof. We prove both claims.

1. It is a plane consequence of formulas (4.13) and (4.17).
2. In order to write down the Hamiltonian vector field v_H, for $H \in \mathscr{F}_{\mathbf{G}}(\mathfrak{g}^*)$, take any $F \in \mathscr{F}(\mathfrak{g}^*)$ and let $\mathsf{X}^* \in \mathfrak{g}^*$. By definition of v_H we have:
$$\{F, H\}_R(\mathsf{X}^*) = \langle\, \mathsf{v}_H(\mathsf{X}^*), \mathrm{d}F(\mathsf{X}^*)\,\rangle. \tag{4.21}$$
From (4.13) and (4.17) we get
$$\begin{aligned} \{F, H\}_R(\mathsf{X}^*) &= \frac{1}{2}\langle \mathsf{X}^*, [\,\mathrm{d}F(\mathsf{X}^*), R(\mathrm{d}H(\mathsf{X}^*))\,]\rangle \\ &= \frac{1}{2}\left\langle\, \mathrm{ad}^*_{R(\mathrm{d}H(\mathsf{X}^*))} \mathsf{X}^*, \mathrm{d}F(\mathsf{X}^*)\,\right\rangle, \end{aligned} \tag{4.22}$$

where in passing from the first to the second line we used (4.16). Comparing (4.21) and (4.22) we get

$$\mathsf{v}_H(\mathsf{X}^*) = \frac{1}{2}\mathrm{ad}^*_{R(\mathrm{d}H(\mathsf{X}^*))}\mathsf{X}^*,$$

which is the desired formula. If we now assume that $\mathfrak{g}$ comes from a Lie algebra splitting and $R := \pi_+ - \pi_-$ we get

$$\begin{aligned} R(\mathrm{d}H(\mathsf{X}^*)) &= (\mathrm{d}H(\mathsf{X}^*))_+ - (\mathrm{d}H(\mathsf{X}^*))_- \\ &= 2(\mathrm{d}H(\mathsf{X}^*))_+ - \mathrm{d}H(\mathsf{X}^*) \\ &= \mathrm{d}H(\mathsf{X}^*) - 2(\mathrm{d}H(\mathsf{X}^*))_-. \end{aligned}$$

So formula (4.20) is easily obtained if we use again (4.17).

The Theorem is proved. ■

► Remarks:

- Theorem 4.3 has a transparent geometrical meaning: it shows that the trajectories of dynamical systems with Hamiltonians $H \in \mathscr{F}_{\mathbf{G}}(\mathfrak{g}^*)$ lie in the intersection of two families of orbits in $\mathfrak{g}^*$, the coadjoint orbits of $(\mathfrak{g}, [\cdot,\cdot])$ and $(\mathfrak{g}, [\cdot,\cdot]_R)$. Indeed, the coadjoint orbits of $(\mathfrak{g}, [\cdot,\cdot]_R)$ are preserved by all Hamiltonian flows in $\mathfrak{g}^*$. On the other hand, formula (4.19) shows that the flow of v_H is always tangent to the $\mathfrak{g}$-orbits in $\mathfrak{g}^*$. In many cases the intersections of orbits are precisely the Liouville tori for our dynamical systems.
- One should not confuse the content of Theorem 4.3 with the bi-Hamiltonian construction of completely integrable systems of Theorem 3.1, where the same equations are Hamiltonian w.r.t. different Poisson structures. In general, the equations of motion induced by Casimirs w.r.t. the R-bracket are not Hamiltonian w.r.t. the Lie Poisson structure on $\mathfrak{g}^*$.

► We now assume that $\mathfrak{g}$ admits a splitting $\mathfrak{g} = \mathfrak{g}_+ \oplus \mathfrak{g}_-$, so that $\mathfrak{g}$ becomes a double Lie algebra with $R := \pi_+ - \pi_-$ in view of Theorem 4.2. We denote by $\mathbf{G}_\pm$ the subgroups of $\mathbf{G}$ corresponding to $\mathfrak{g}_\pm$. The next claim shows that the vector field (4.20) can be explicitly integrated by factorization.

Theorem 4.4 (*Adler-Kostant-Symes*)

Let $H \in \mathscr{F}_{\mathbf{G}}(\mathfrak{g}^*)$. Let $\mathfrak{g} = \mathfrak{g}_+ \oplus \mathfrak{g}_-$ be a double Lie algebra with $R := \pi_+ - \pi_-$. For $|t|$ small, let $t \mapsto h_\pm(t)$ denote the smooth curves in $\mathbf{G}_\pm$ which solve the factorization problem

$$\exp(-t\,\mathrm{d}H(\mathsf{X}_0^*)) = h_+^{-1}(t)\,h_-(t), \qquad h_\pm(0) = e, \tag{4.23}$$

where $\mathsf{X}_0^* := \mathsf{X}^*(0) \in \mathfrak{g}^*$*. Then the integral curve of (4.20) which starts at* X_0^* *is given by*

$$\mathsf{X}^*(t) = \mathsf{Ad}^*_{h_+(t)}\,\mathsf{X}_0^* = \mathsf{Ad}^*_{h_-(t)}\,\mathsf{X}_0^*.$$

Proof. We proceed by steps.

- We first show that

$$\mathsf{Ad}^*_{h_+(t)}\,\mathsf{X}_0^* = \mathsf{Ad}^*_{h_-(t)}\,\mathsf{X}_0^*.$$

  Since Ad^* is a group homomorphism, the factorization (4.23) implies that

$$\mathsf{Ad}^*_{\exp(-t\,\mathsf{d}H(\mathsf{X}_0^*))}\,\mathsf{X}_0^* = \mathsf{Ad}^*_{h_+^{-1}(t)}\,\mathsf{Ad}^*_{h_-(t)}\,\mathsf{X}_0^*,$$

  for any $\mathsf{X}_0^* \in \mathfrak{g}^*$. For any $\mathsf{X} \in \mathfrak{g}$ we have:

$$\begin{aligned}\left\langle \mathsf{Ad}^*_{\exp(-t\,\mathsf{d}H(\mathsf{X}_0^*))}\,\mathsf{X}_0^*, \mathsf{X} \right\rangle &= \left\langle \mathsf{X}_0^*, \mathsf{Ad}_{\exp(t\,\mathsf{d}H(\mathsf{X}_0^*))}\,\mathsf{X} \right\rangle \\ &= \left\langle \mathsf{X}_0^*, \exp\left(t\,\mathsf{ad}_{\mathsf{d}H(\mathsf{X}_0^*)}\right)\mathsf{X} \right\rangle \\ &= \langle \mathsf{X}_0^*, \mathsf{X} \rangle + t\,\langle \mathsf{X}_0^*, [\,\mathsf{d}H(\mathsf{X}_0^*), \star\,] \rangle \\ &= \langle \mathsf{X}_0^*, \mathsf{X} \rangle,\end{aligned}$$

  where the value of $\star$, which depends on t, is irrelevant because $H \in \mathscr{F}_{\mathsf{G}}(\mathfrak{g}^*)$ (see formula (4.17)). Note that we used (1.37) is passing from the first to the second line. Therefore we obtained

$$\mathsf{Ad}^*_{\exp(-t\,\mathsf{d}H(\mathsf{X}_0^*))}\,\mathsf{X}_0^* = \mathsf{X}_0^*$$

  and hence that $\mathsf{Ad}^*_{h_+(t)}\,\mathsf{X}_0^* = \mathsf{Ad}^*_{h_-(t)}\,\mathsf{X}_0^*$.

- For small $|t|$ we define $\mathsf{X}^*(t) := \mathsf{Ad}^*_{h_+(t)}\,\mathsf{X}_0^*$ and we show that it solves (4.20) with $+$. Equivalently we want to show that

$$\frac{\mathsf{d}}{\mathsf{d}t}\left(\mathsf{Ad}^*_{h_+(t)}\,\mathsf{X}_0^*\right) = \mathsf{ad}^*_{(\mathsf{d}H(\mathsf{X}^*(t)))_+}\,\mathsf{X}^*(t). \tag{4.24}$$

- Recall that Ad and Ad^* are just given by conjugation. Then, the l.h.s. of (4.24) is

$$\frac{\mathsf{d}}{\mathsf{d}t}\left(\mathsf{Ad}^*_{h_+(t)}\,\mathsf{X}_0^*\right) = \mathsf{ad}^*_{\dot{h}_+(t)\,h_+^{-1}(t)}\,\mathsf{X}^*(t),$$

  so it suffices to show that

$$\dot{h}_+(t)\,h_+^{-1}(t) = (\mathsf{d}H(\mathsf{X}^*(t)))_+. \tag{4.25}$$

- We write (4.23) as
$$h_+(t)\exp(-t\,\mathrm{d}H(\mathsf{X}_0^*)) = h_-(t), \tag{4.26}$$
and we differentiate w.r.t. t:
$$\dot{h}_+(t)\exp(-t\,\mathrm{d}H(\mathsf{X}_0^*)) - h_+(t)\exp(-t\,\mathrm{d}H(\mathsf{X}_0^*))\,\mathrm{d}H(\mathsf{X}_0^*) = \dot{h}_-(t).$$

- Multiply both sides of the last equation by $h_-^{-1}(t)$,
$$\begin{aligned}&\dot{h}_+(t)\exp(-t\,\mathrm{d}H(\mathsf{X}_0^*))\,h_-^{-1}(t)\\ &\qquad - h_+(t)\exp(-t\,\mathrm{d}H(\mathsf{X}_0^*))\,\mathrm{d}H(\mathsf{X}_0^*)h_-^{-1}(t) = \dot{h}_-(t)\,h_-^{-1}(t),\end{aligned}$$
that is (use (4.26))
$$\dot{h}_+(t)\,h_+^{-1}(t) - h_-(t)\,\mathrm{d}H(\mathsf{X}_0^*)\,h_-^{-1}(t) = \dot{h}_-(t)\,h_-^{-1}(t). \tag{4.27}$$

- Now, the Ad^*-invariance of H implies formula (4.18), so that
$$\mathrm{d}H(\mathsf{X}^*(t)) = \mathrm{d}H\left(\mathrm{Ad}^*_{h_-(t)}\,\mathsf{X}_0^*\right) = \mathrm{Ad}^*_{h_-(t)}\,\mathrm{d}H(\mathsf{X}_0^*) = h_-(t)\mathrm{d}H(\mathsf{X}_0^*)h_-^{-1}(t),$$
so that (4.27) becomes
$$\dot{h}_+(t)h_+^{-1}(t) - \mathrm{d}H(\mathsf{X}^*(t)) = \dot{h}_-(t)\,h_-^{-1}(t).$$

- If we take the $+$ part of both sides of the last equation we find
$$\dot{h}_+(t)\,h_+^{-1}(t) = (\mathrm{d}H(\mathsf{X}^*(t)))_+,$$
that is exactly (4.25).

The Theorem is proved. ■

► As for the derivation of the Lie-Poisson manifold $(\mathfrak{g}, \{\cdot,\cdot\}_{\mathrm{LP}})$ from $(\mathfrak{g}^*, \{\cdot,\cdot\}_{\mathrm{LP}})$, assuming that $\mathfrak{g}$ is a quadratic Lie algebra, we can now construct a linear R-bracket on $\mathfrak{g}$. This allows us to transcribe Theorems 4.3 and 4.4 to $\mathfrak{g}$. According to definition 4.2 we can define the following *linear R-bracket* on $\mathfrak{g}$:
$$\begin{aligned}\{F,G\}_R(\mathsf{X}) &:= \frac{1}{2}\langle \mathsf{X} \,|\, [R(\nabla F(\mathsf{X})), \nabla G(\mathsf{X})]\rangle \\ &+ \frac{1}{2}\langle \mathsf{X} \,|\, [\nabla F(\mathsf{X}), R(\nabla G(\mathsf{X}))]\rangle.\end{aligned} \tag{4.28}$$

► We now give a formulation of Theorems 4.3 and 4.4 on $\mathfrak{g}$. They are more suitable for applications because they allow to write down Hamiltonian vector fields in Lax form.

Theorem 4.5 (*Semenov-Tian-Shansky*)

Let $(\mathfrak{g}, [\cdot,\cdot], [\cdot,\cdot]_R)$ be a double Lie algebra which is also quadratic. Then:

1. *Any two functions $F, H \in \mathscr{F}_{\mathbf{G}}(\mathfrak{g})$ are in involution w.r.t. the Lie-Poisson R-bracket (4.28), i.e., $\{F, H\}_R = 0$.*
2. *Let $H \in \mathscr{F}_{\mathbf{G}}(\mathfrak{g})$. The Hamiltonian vector field $\mathsf{v}_H := \{\cdot, H\}_R$ takes the form*

$$\mathsf{v}_H(\mathsf{X}) = -\frac{1}{2}\,[\mathsf{X},\mathsf{B}], \qquad \mathsf{B} := R(\nabla H(\mathsf{X})). \tag{4.29}$$

In particular, if $\mathfrak{g} = \mathfrak{g}_+ \oplus \mathfrak{g}_-$ and $R := \pi_+ - \pi_-$, one has

$$\mathsf{v}_H(\mathsf{X}) = \pm\,[\mathsf{X},\mathsf{B}_\mp], \qquad \mathsf{B}_\pm := (\nabla H(\mathsf{X}))_\pm. \tag{4.30}$$

No Proof.

Theorem 4.6 (*Adler-Kostant-Symes*)

Let $H \in \mathscr{F}_{\mathbf{G}}(\mathfrak{g})$. Let $(\mathfrak{g}, [\cdot,\cdot], [\cdot,\cdot]_R)$ be a double Lie algebra, with $R := \pi_+ - \pi_-$, which is also quadratic and assume that $\mathfrak{g} = \mathfrak{g}_+ \oplus \mathfrak{g}_-$. For $|t|$ small, let $t \mapsto h_\pm(t)$ denote the smooth curves in $\mathbf{G}_\pm$ which solve the factorization problem

$$\exp(-t\,\nabla H(\mathsf{X}_0)) = h_+^{-1}(t)\,h_-(t), \qquad h_\pm(0) = e, \tag{4.31}$$

where $\mathsf{X}(0) := \mathsf{X}_0 \in \mathfrak{g}$. Then the integral curve of (4.30) which starts at X_0 is given by

$$\mathsf{X}(t) = \mathsf{Ad}_{h_+(t)}\,\mathsf{X}_0 = \mathsf{Ad}_{h_-(t)}\,\mathsf{X}_0.$$

No Proof.

▶ Remarks:

- The presented R-matrix approach shows that there exists an algebraic procedure which allows one to construct Hamiltonian flows on $\mathfrak{g}^*$ or $\mathfrak{g}$ which are explicitly solvable. However, the question whether these Hamiltonian equations are completely integrable in the sense of Definition 3.1 requires a separate study.
- Indeed, the integrals of motion of (4.29) come from the Casimirs of $\mathfrak{g}$. If $\mathbf{G}$ is a finite-dimensional semi-simple Lie group, the ring $\mathscr{F}_{\mathbf{G}}(\mathfrak{g})$ of its coadjoint invariants is finitely generated. A general theorem says that the number of independent generators of this ring is equal to $\mathsf{rank}\,\mathfrak{g} \leqslant \mathsf{dim}\,\mathfrak{g}$ (note that the rank of $\mathfrak{g}$ coincides with the rank of the Lie-Poisson structure defined on $\mathfrak{g}$). It turns out that for typical orbits the number of Poisson involutive and independent integrals of motion provided by the R-matrix approach is insufficient.

- For generic Hamiltonian systems this would mean that their trajectory curves densely span a subset of small codimension in the phase space. In practice, however, the situation is exactly opposite: the trajectories of the Hamiltonian flows generated by (4.29) span a subset of small dimension. This means that the behavior of our exactly solvable Lax equations (4.29) is much more regular than for typical Hamiltonian systems.
- However, this regular behaviour also prevents us from constructing immediately a complete set of integrals of motion in Poisson involution. The situation changes drastically for Lax equations with spectral parameter. In this case the ring of invariants of the underlying loop Lie algebra has infinitely many independent generators. Accordingly, Lax equations supported on generic finite-dimensional orbits of the R-bracket are automatically completely integrable in the Arnold-Liouville sense (at least, under some mild technical assumptions).

4.2.4 Lax hierarchies

► As a matter of fact, integrable systems appear not separately, but are organized in hierarchies. This claim is already evident in Theorem 4.5: for each Ad-invariant function we can generate a vector field which admits a Lax equation of type (4.29). One can even show that such Lax equations admit a whole hierarchy of compatible Poisson structure.

► Consider again the framework of Theorem 4.5.

- Let $F_1, \dots, F_s$ be s Ad-invariant functions, i.e., $F_i \in \mathscr{F}_{\mathrm{G}}(\mathfrak{g})$, $i = 1, \dots, s$. Set $L_i := (\nabla F_i)/2$.
- Consider the vector fields $\mathsf{v}_{F_1}, \dots, \mathsf{v}_{F_s}$ on $\mathfrak{g}$ defined by

$$\frac{\mathsf{d}}{\mathsf{d}t_i}\mathsf{X} = \mathsf{v}_{F_i}(\mathsf{X}) = [\,R(L_i(\mathsf{X})), \mathsf{X}\,], \qquad \mathsf{X} \in \mathfrak{g}, \tag{4.32}$$

  with $i = 1, \dots, s$.

Theorem 4.7

If

$$\frac{\mathsf{d}}{\mathsf{d}t_i}(R\,\mathsf{X}) = R\frac{\mathsf{d}}{\mathsf{d}t_i}\mathsf{X} \tag{4.33}$$

for any $\mathsf{X} \in \mathfrak{g}$ *and for all* $i = 1, \dots, s$, *and the modified Yang-Baxter equation (4.11) is satisfied, then the vector fields* $\mathsf{v}_{F_1}, \dots, \mathsf{v}_{F_s}$ *are pairwise commuting.*

Proof. We proceed by steps.

- Consider any two arbitrary vector fields $\mathsf{v}_{F_j}, \mathsf{v}_{F_k}$ and define

$$\Delta := \frac{\mathrm{d}}{\mathrm{d}t_k}\left(\frac{\mathrm{d}}{\mathrm{d}t_j}\mathsf{X}\right) - \frac{\mathrm{d}}{\mathrm{d}t_j}\left(\frac{\mathrm{d}}{\mathrm{d}t_k}\mathsf{X}\right).$$

  We want to prove that $\Delta = 0$ for all $\mathsf{X} \in \mathfrak{g}$.

- We have:

$$\begin{aligned}
\Delta &= \frac{\mathrm{d}}{\mathrm{d}t_k}\left[R(L_j(\mathsf{X})), \mathsf{X}\right] - \frac{\mathrm{d}}{\mathrm{d}t_j}\left[R(L_k(\mathsf{X})), \mathsf{X}\right] \\
&= \left[\frac{\mathrm{d}}{\mathrm{d}t_k}\left(R(L_j(\mathsf{X}))\right) - \frac{\mathrm{d}}{\mathrm{d}t_j}\left(R(L_k(\mathsf{X}))\right), \mathsf{X}\right] + \left[R(L_j(\mathsf{X})), \left[R(L_k(\mathsf{X})), \mathsf{X}\right]\right] \\
&\quad - \left[R(L_k(\mathsf{X})), \left[R(L_j(\mathsf{X})), \mathsf{X}\right]\right] \\
&= \left[\frac{\mathrm{d}}{\mathrm{d}t_k}\left(R(L_j(\mathsf{X}))\right) - \frac{\mathrm{d}}{\mathrm{d}t_j}\left(R(L_k(\mathsf{X}))\right), \mathsf{X}\right] + \left[\left[R(L_j(\mathsf{X})), R(L_k(\mathsf{X}))\right], \mathsf{X}\right],
\end{aligned}$$

  where we used the Jacobi identity on $\mathfrak{g}$.

- If condition (4.33) holds true then

$$\Delta = \left[R\frac{\mathrm{d}}{\mathrm{d}t_k}(L_j(\mathsf{X})) - R\frac{\mathrm{d}}{\mathrm{d}t_j}(L_k(\mathsf{X})), \mathsf{X}\right] + \left[\left[R(L_j(\mathsf{X})), R(L_k(\mathsf{X}))\right], \mathsf{X}\right].$$

- Note that

$$\frac{\mathrm{d}}{\mathrm{d}t_k}(L_j(\mathsf{X})) = \left[R(L_k(\mathsf{X})), L_j(\mathsf{X})\right]$$

  is nothing but the directional derivative of L_j along the direction of (4.32) with $i = k$.

- Therefore we get

$$\begin{aligned}
\Delta &= \left[R\left(\left[R(L_k(\mathsf{X})), L_j(\mathsf{X})\right]\right) - R\left(\left[R(L_j(\mathsf{X})), L_k(\mathsf{X})\right]\right), \mathsf{X}\right] \\
&\quad + \left[\left[R(L_j(\mathsf{X})), R(L_k(\mathsf{X}))\right], \mathsf{X}\right] \\
&= -\left[Q_R(L_j(\mathsf{X}), L_k(\mathsf{X})), \mathsf{X}\right],
\end{aligned}$$

  where we used (4.7).

- If the modified Yang-Baxter equation (4.11) is satisfied then

$$\Delta = c\left[\left[L_j(\mathsf{X}), L_k(\mathsf{X})\right], \mathsf{X}\right],$$

  which vanishes because the functions F_j and F_k are Ad-invariant.

The Theorem is proved. ■

4.3 r-brackets

► We now describe a more general approach to Lax equations, where the essential ingredient is a Lie algebra structure on $\mathfrak{g}^*$ (which yields a linear Poisson structure on $\mathfrak{g}$). We shall study Lie algebras $\mathfrak{g}$ (not necessarily quadratic) whose dual vector space $\mathfrak{g}^*$ carries a Lie algebra structure satisfying a suitable compatibility condition, with that of $\mathfrak{g}$ itself. Such objects are called *Lie bialgebras*. The construction which follows can be done also for loop algebras.

4.3.1 *Lie bialgebras and coboundary Lie bialgebras*

► Let us recall some notions introduced in Chapter 1.

- Adjoint actions can be generalized to the *tensor algebra* of $\mathfrak{g}$:

$$\mathscr{T}(\mathfrak{g}) := \bigoplus_{k\geqslant 0} \mathfrak{g}^{\otimes k}, \qquad \mathfrak{g}^{\otimes k} := \underbrace{\mathfrak{g}\otimes\cdots\otimes\mathfrak{g}}_{k\,\text{times}}.$$

(a) Let us fix $k = 2$. Consider $\mathsf{X},\mathsf{Y},\mathsf{Z} \in \mathfrak{g}$. Then

$$\mathsf{ad}^{(2)}_{\mathsf{X}}(\mathsf{Y}\otimes\mathsf{Z}) := (\mathsf{ad}_{\mathsf{X}}\mathsf{Y})\otimes\mathsf{Z} + \mathsf{Y}\otimes(\mathsf{ad}_{\mathsf{X}}\mathsf{Z}) = [\mathsf{X},\mathsf{Y}]\otimes\mathsf{Z} + \mathsf{Y}\otimes[\mathsf{X},\mathsf{Z}].$$

If 1 denotes the identity map from $\mathfrak{g}$ to $\mathfrak{g}$ (note that we are embedding $\mathfrak{g}$ into an associative algebra with unity) we write

$$\mathsf{ad}^{(2)}_{\mathsf{X}} = \mathsf{ad}_{\mathsf{X}}\otimes 1 + 1\otimes\mathsf{ad}_{\mathsf{X}}.$$

The map $\mathsf{ad}^{(2)} : \mathfrak{g} \longrightarrow \mathsf{End}(\mathfrak{g}\otimes\mathfrak{g})$ provides a representation of $\mathfrak{g}$ on $\mathfrak{g}\otimes\mathfrak{g}$.

(b) Taking into account the definition of the coadjoint action of $\mathfrak{g}$,

$$\langle\,\mathsf{X}^*,\mathsf{ad}_{\mathsf{X}}\,\mathsf{Y}\,\rangle = -\,\langle\,\mathsf{ad}^*_{\mathsf{X}}\,\mathsf{X}^*,\mathsf{Y}\,\rangle, \qquad \mathsf{X}^*\in\mathfrak{g}^*,\mathsf{X},\mathsf{Y}\in\mathfrak{g},$$

one can also introduce a coadjoint action for tensor products. For example,

$$\langle\,\mathsf{X}^*\otimes\mathsf{Y}^*,\mathsf{ad}_{\mathsf{X}}\,\mathsf{Y}\otimes\mathsf{Z}\,\rangle = -\,\langle\,\mathsf{ad}^*_{\mathsf{X}}\,\mathsf{X}^*\otimes\mathsf{Y}^*,\mathsf{Y}\otimes\mathsf{Z}\,\rangle, \tag{4.34}$$

for all $\mathsf{X}^*,\mathsf{Y}^* \in \mathfrak{g}^*,\mathsf{X},\mathsf{Y},\mathsf{Z}\in\mathfrak{g}$. Here $\langle\,\cdot,\cdot\,\rangle$ is the natural pairing between $\mathscr{T}(\mathfrak{g})$ and $\mathscr{T}(\mathfrak{g}^*)$.

(c) From (4.34) there follows a useful formula valid for the map $\mathsf{ad}^{(2)}$. If $r \in \mathfrak{g}\otimes\mathfrak{g}$ one has

$$\left\langle\,\mathsf{X}^*\otimes\mathsf{Y}^*,\mathsf{ad}^{(2)}_{\mathsf{X}}\,r\,\right\rangle = -\,\langle\,r,\mathsf{ad}^*_{\mathsf{X}}\,\mathsf{X}^*\otimes\mathsf{Y}^* + \mathsf{X}^*\otimes\mathsf{ad}^*_{\mathsf{X}}\,\mathsf{Y}^*\,\rangle, \tag{4.35}$$

for all $\mathsf{X}^*,\mathsf{Y}^* \in \mathfrak{g}^*,\mathsf{X}\in\mathfrak{g}$.

- Consider the following space of ℓ-cochains of $\mathfrak{g}$:

$$\mathcal{C}^\ell(\mathfrak{g},\mathfrak{g}\otimes\mathfrak{g}) := \mathsf{Hom}\left(\mathfrak{g}^{\wedge\ell},\mathfrak{g}\otimes\mathfrak{g}\right).$$

Then $\mathcal{C}^\ell(\mathfrak{g},\mathfrak{g}\otimes\mathfrak{g})$ is equipped with a coboundary operator $\delta : \mathcal{C}^\ell(\mathfrak{g},\mathfrak{g}\otimes\mathfrak{g}) \to \mathcal{C}^{\ell+1}(\mathfrak{g},\mathfrak{g}\otimes\mathfrak{g})$.

(a) Let $r \in \mathfrak{g}\otimes\mathfrak{g} \simeq \mathcal{C}^0(\mathfrak{g},\mathfrak{g}\otimes\mathfrak{g})$. Then we can define a linear map $\gamma := \delta r : \mathfrak{g} \to \mathfrak{g}\otimes\mathfrak{g}$ given by (see (1.44))

$$\gamma(\mathsf{X}) := \delta r(\mathsf{X}) = \mathsf{ad}^{(2)}_{\mathsf{X}} r \tag{4.36}$$

for all $\mathsf{X} \in \mathfrak{g}$.

(b) The transpose map of γ is the linear map $\gamma^\top : \mathfrak{g}^*\otimes\mathfrak{g}^* \to \mathfrak{g}^*$ which can be viewed as a bilinear map $\mathfrak{g}^*\times\mathfrak{g}^* \to \mathfrak{g}^*$, still denoted by $\gamma^\top$. Explicitly we can write

$$\left\langle \gamma^\top(\mathsf{X}^*,\mathsf{Y}^*),\mathsf{Z} \right\rangle = (\mathsf{X}^*\otimes\mathsf{Y}^*)\,\mathsf{ad}^{(2)}_{\mathsf{Z}} r \qquad \forall \mathsf{X}^*,\mathsf{Y}^* \in \mathfrak{g}^*,\ \mathsf{Z}\in\mathfrak{g}.$$

► We will be interested in the case when $\gamma^\top$ defines a Lie algebra structure on $\mathfrak{g}^*$. It is therefore essential to give the following definition.

Definition 4.4

1. *A* **Lie bialgebra** *is a Lie algebra* $(\mathfrak{g},[\,\cdot,\cdot\,])$ *equipped with a linear map* $\gamma : \mathfrak{g} \to \mathfrak{g}\otimes\mathfrak{g}$ *(called* **cobracket***) such that*

   (a) The bracket $[\,\cdot,\cdot\,]_\gamma$ *on* $\mathfrak{g}^*$ *defined by*

$$[\,\mathsf{X}^*,\mathsf{Y}^*\,]_\gamma := \gamma^\top(\mathsf{X}^*\otimes\mathsf{Y}^*),$$

   or, equivalently,

$$\langle\,[\,\mathsf{X}^*,\mathsf{Y}^*\,]_\gamma,\mathsf{X}\,\rangle := \langle\,\gamma(\mathsf{X}),\mathsf{X}^*\otimes\mathsf{Y}^*\,\rangle,$$

   is a Lie bracket for all $\mathsf{X}^*,\mathsf{Y}^* \in \mathfrak{g}^*, \mathsf{X}\in\mathfrak{g}$.

   (b) γ *is a 1-cocycle on* $\mathfrak{g}$ *with values on* $\mathfrak{g}\otimes\mathfrak{g}$*, where* $\mathfrak{g}$ *acts on* $\mathfrak{g}\otimes\mathfrak{g}$ *by means of* $\mathsf{ad}^{(2)}$*, i.e.,*

$$\mathsf{ad}^{(2)}_{\mathsf{X}}(\gamma(\mathsf{Y})) - \mathsf{ad}^{(2)}_{\mathsf{Y}}(\gamma(\mathsf{X})) - \gamma([\mathsf{X},\mathsf{Y}]) = 0,$$

   for all $\mathsf{X},\mathsf{Y}\in\mathfrak{g}$.

   A Lie bialgebra is denoted by $(\mathfrak{g},[\,\cdot,\cdot\,],[\,\cdot,\cdot\,]_\gamma)$.

2. *A Lie bialgebra* $(\mathfrak{g}, [\cdot,\cdot], [\cdot,\cdot]_\gamma)$ *is called* **coboundary Lie bialgebra** *if there exists* $r \in \mathfrak{g}\otimes\mathfrak{g}$ *(called* r**-matrix***) such that* $\gamma = \delta r$*. The corresponding Lie bracket on* $\mathfrak{g}^*$ *is*

$$[\mathsf{X}^*,\mathsf{Y}^*]_r := (\delta r)^\top(\mathsf{X}^*\otimes\mathsf{Y}^*), \qquad \forall\,\mathsf{X}^*,\mathsf{Y}^*\in\mathfrak{g}^*, \tag{4.37}$$

or, equivalently,

$$\langle\,[\mathsf{X}^*,\mathsf{Y}^*]_r,\mathsf{X}\,\rangle := \langle\,\delta r(\mathsf{X}),\mathsf{X}^*\otimes\mathsf{Y}^*\,\rangle \qquad \forall\,\mathsf{X}^*,\mathsf{Y}^*\in\mathfrak{g}^*,\mathsf{X}\in\mathfrak{g}. \tag{4.38}$$

A coboundary Lie bialgebra is denoted by $(\mathfrak{g}, [\cdot,\cdot], [\cdot,\cdot]_r)$.

▶ Notational remarks:

- If we view the $(\delta r)^\top$ as a bilinear map $\mathfrak{g}^*\times\mathfrak{g}^*\to\mathfrak{g}^*$, still denoted by $(\delta r)^\top$, we can write (4.37) as

$$[\mathsf{X}^*,\mathsf{Y}^*]_r := (\delta r)^\top(\mathsf{X}^*,\mathsf{Y}^*), \qquad \forall\,\mathsf{X}^*,\mathsf{Y}^*\in\mathfrak{g}^*.$$

- To emphasize the skew-symmetry of the bracket (4.38) one can also write (4.38) as

$$\langle\,[\mathsf{X}^*,\mathsf{Y}^*]_r,\mathsf{X}\,\rangle := \langle\,\delta r(\mathsf{X}),\mathsf{X}^*\wedge\mathsf{Y}^*\,\rangle \qquad \forall\,\mathsf{X}^*,\mathsf{Y}^*\in\mathfrak{g}^*,\mathsf{X}\in\mathfrak{g}.$$

- Recall that when $\mathsf{Y}\in\mathfrak{g}$ one writes $\mathrm{ad}_\mathsf{X}\mathsf{Y} = [\mathsf{X},\mathsf{Y}]$, with $\mathsf{X}\in\mathfrak{g}$. In the same way, when $r\in\mathfrak{g}\otimes\mathfrak{g}$ one writes

$$\mathrm{ad}^{(2)}_\mathsf{X}\, r = (\mathrm{ad}_\mathsf{X}\otimes 1 + 1\otimes\mathrm{ad}_\mathsf{X})\, r = [\,\mathsf{X}\otimes 1 + 1\otimes\mathsf{X}, r\,],$$

with $\mathsf{X}\in\mathfrak{g}$. Therefore, another equivalent form for the bracket (4.38) is

$$\langle\,[\mathsf{X}^*,\mathsf{Y}^*]_r,\mathsf{X}\,\rangle := \langle\,[\,\mathsf{X}\otimes 1 + 1\otimes\mathsf{X}, r\,],\mathsf{X}^*\wedge\mathsf{Y}^*\,\rangle \qquad \forall\,\mathsf{X}^*,\mathsf{Y}^*\in\mathfrak{g}^*,\mathsf{X}\in\mathfrak{g}.$$

4.3.2 Linear r-brackets and Yang-Baxter equation

▶ Our task is to find necessary and sufficient conditions on $r\in\mathfrak{g}\otimes\mathfrak{g}$ for which $(\mathfrak{g}, [\cdot,\cdot], [\cdot,\cdot]_r)$ is a coboundary Lie bialgebra, i.e., r is an r-matrix.

- First note that a 1-cochain on $\mathfrak{g}$ with values on $\mathfrak{g}\otimes\mathfrak{g}$ which is the coboundary of a 0-cochain on $\mathfrak{g}$ with values on $\mathfrak{g}\otimes\mathfrak{g}$, i.e., of an element $r\in\mathfrak{g}\otimes\mathfrak{g}$, is necessarily a 1-cocycle.

- Therefore we need to satisfy two conditions (for general r the bracket (4.37) is not a Lie bracket on $\mathfrak{g}^*$):

  1. The bracket (4.37) must be skew-symmetric.

2. The bracket (4.37) must satisfy the Jacobi identity.

We immediately notice that the first condition is satisfied if and only if δr take values in the exterior algebra $\mathfrak{g}^{\wedge 2} = \mathfrak{g} \wedge \mathfrak{g}$ (see (1.34)). Necessary and sufficient conditions on r such that the above two conditions be satisfied will be given in Theorem 4.8.

► We need some additional notions and definitions.

- Let $r \in \mathfrak{g} \otimes \mathfrak{g}$. Then r admits a natural splitting into its symmetric part, say r_+, and skew-symmetric part, say r_-:

$$r = r_+ + r_-, \qquad r_+ \in \mathscr{S}^2(\mathfrak{g}),\, r_- \in \mathfrak{g}^{\wedge 2}.$$

Here $\mathscr{S}^2(\mathfrak{g})$ is the symmetric algebra obtained as the quotient algebra of $\mathfrak{g} \otimes \mathfrak{g}$ by the two-sided ideal generated by all elements of the form $\mathsf{X} \otimes \mathsf{Y} - \mathsf{Y} \otimes \mathsf{X}$ with $\mathsf{X}, \mathsf{Y} \in \mathfrak{g}$. Note that the linearity of δ implies

$$\delta r(\mathsf{X}) = \delta r_+(\mathsf{X}) + \delta r_-(\mathsf{X}) \qquad \forall\, \mathsf{X} \in \mathfrak{g}. \tag{4.39}$$

- We introduce a linear map $\varrho : \mathfrak{g}^* \to \mathfrak{g}$ defined as

$$\langle\, \mathsf{X}^*, \varrho(\mathsf{Y}^*)\,\rangle := \langle\, r, \mathsf{X}^* \otimes \mathsf{Y}^*\,\rangle,$$

for any $\mathsf{X}^*, \mathsf{Y}^* \in \mathfrak{g}^*$. Let $\varrho^\top : \mathfrak{g}^* \to \mathfrak{g}$ denote the transpose map of ϱ. Then we define

$$\varrho_\pm := \frac{1}{2}\left(\varrho \pm \varrho^\top\right).$$

- Assume that $r \in \mathfrak{g}^{\wedge 2}$. Then the *algebraic Schouten bracket* $[\![r, r]\!]$ is an element of $\mathfrak{g}^{\wedge 3}$ defined by

$$[\![r, r]\!](\mathsf{X}^*, \mathsf{Y}^*, \mathsf{Z}^*) := 2\,\langle\, [\,\varrho(\mathsf{X}^*), \varrho(\mathsf{Y}^*)\,], \mathsf{Z}^*\,\rangle + \circlearrowleft (\mathsf{X}^*, \mathsf{Y}^*, \mathsf{Z}^*),$$

with $\mathsf{X}^*, \mathsf{Y}^*, \mathsf{Z}^* \in \mathfrak{g}^*$.

► We now prove some intermediate results. The next Lemma gives conditions such that the bracket (4.37) is skew-symmetric.

Lemma 4.1

Let $(\mathfrak{g}, [\,\cdot,\cdot\,])$ be a Lie algebra and let $r = r_+ + r_-, r_+ \in \mathscr{S}^2(\mathfrak{g}),\, r_- \in \mathfrak{g}^{\wedge 2}$.

1. *For any $\mathsf{X} \in \mathfrak{g}$ we have $\delta r_+(\mathsf{X}) \in \mathscr{S}^2(\mathfrak{g})$ and $\delta r_-(\mathsf{X}) \in \mathfrak{g}^{\wedge 2}$.*
2. *The bracket (4.37) is skew-symmetric if and only if $\delta r_+ = 0$, i.e., $\delta r = \delta r_-$.*

Proof. We prove both claims.

1. We prove only that $\delta r_-(\mathsf{X}) \in \mathfrak{g}^{\wedge 2}$. The second claim can be proved in a similar way. Let $\mathsf{X} \in \mathfrak{g}$ and $\mathsf{X}^*, \mathsf{Y}^* \in \mathfrak{g}^*$. Then by using (4.35) and the skew-symmetry of r_- we get

$$\begin{aligned}\langle \delta r_-(\mathsf{X}), \mathsf{X}^* \otimes \mathsf{Y}^* \rangle &= \left\langle \mathrm{ad}^{(2)}_{\mathsf{X}} r_-, \mathsf{X}^* \otimes \mathsf{Y}^* \right\rangle \\ &= -\langle r_-, \mathrm{ad}^*_{\mathsf{X}} \mathsf{X}^* \otimes \mathsf{Y}^* + \mathsf{X}^* \otimes \mathrm{ad}^*_{\mathsf{X}} \mathsf{Y}^* \rangle \\ &= \langle r_-, \mathsf{Y}^* \otimes \mathrm{ad}^*_{\mathsf{X}} \mathsf{X}^* + \mathrm{ad}^*_{\mathsf{X}} \mathsf{Y}^* \otimes \mathsf{X}^* \rangle \\ &= -\left\langle \mathrm{ad}^{(2)}_{\mathsf{X}} r_-, \mathsf{Y}^* \otimes \mathsf{X}^* \right\rangle \\ &= -\langle \delta r_-(\mathsf{X}), \mathsf{Y}^* \otimes \mathsf{X}^* \rangle,\end{aligned}$$

   from which there follows that $\delta r_-(\mathsf{X}) \in \mathfrak{g}^{\wedge 2}$.

2. This follows from definition (4.37), from (4.39) and from the first claim of the Lemma. Indeed we have:

$$\langle [\mathsf{X}^*, \mathsf{Y}^*]_r, \mathsf{X} \rangle = -\langle [\mathsf{Y}^*, \mathsf{X}^*]_r, \mathsf{X} \rangle \qquad \forall \mathsf{X}^*, \mathsf{Y}^* \in \mathfrak{g}^*, \, \mathsf{X} \in \mathfrak{g}$$

   if and only if $\delta r_+ = 0$.

The Lemma is proved. ∎

► In view of Lemma 4.1 we assume in the next claim that r is skew-symmetric. The next Lemma gives conditions such that the bracket (4.37) satisfies the Jacobi identity.

Lemma 4.2

Let $(\mathfrak{g}, [\cdot,\cdot])$ be a Lie algebra and let $r \in \mathfrak{g}^{\wedge 2}$.

1. *The (skew-symmetric) bracket (4.37) takes the form*

$$[\mathsf{X}^*, \mathsf{Y}^*]_r = \mathrm{ad}^*_{\varrho(\mathsf{X}^*)} \mathsf{Y}^* - \mathrm{ad}^*_{\varrho(\mathsf{Y}^*)} \mathsf{X}^*, \tag{4.40}$$

   for all $\mathsf{X}^*, \mathsf{Y}^*$.

2. *There holds*

$$\langle [[\mathsf{X}^*, \mathsf{Y}^*]_r, \mathsf{Z}^*]_r, \mathsf{X} \rangle + \circlearrowleft (\mathsf{X}^*, \mathsf{Y}^*, \mathsf{Z}^*) = -\frac{1}{2} \delta [\![r, r]\!] (\mathsf{X}^*, \mathsf{Y}^*, \mathsf{Z}^*)(\mathsf{X}),$$

   for all $\mathsf{X}^*, \mathsf{Y}^*, \mathsf{Z}^* \in \mathfrak{g}^*$ and $\mathsf{X} \in \mathfrak{g}$.

3. *The bracket (4.40) satisfies the Jacobi identity if and only if $\delta [\![r, r]\!] = 0$.*

Proof. We prove only the first claim. The second one requires a quite long computation, while the third one is a plane consequence of the second claim. For $X^*, Y^* \in \mathfrak{g}^*$, $X \in \mathfrak{g}$ we have

$$\begin{aligned}
\langle [X^*, Y^*]_r, X \rangle &:= \langle \delta r(X), X^* \otimes Y^* \rangle = \left\langle \mathrm{ad}_X^{(2)} r, X^* \otimes Y^* \right\rangle \\
&= -\langle r, \mathrm{ad}_X^* X^* \otimes Y^* + X^* \otimes \mathrm{ad}_X^* Y^* \rangle \\
&= -\langle \mathrm{ad}_X^* X^*, \varrho(Y^*) \rangle + \langle \mathrm{ad}_X^* Y^*, \varrho(X^*) \rangle \\
&= \left\langle \mathrm{ad}_{\varrho(Y^*)}^* X^*, X \right\rangle - \left\langle \mathrm{ad}_{\varrho(X^*)}^* Y^*, X \right\rangle,
\end{aligned}$$

which is the desired formula. ■

► The next Theorem is now a consequence of Lemmas 4.1 and 4.2.

Theorem 4.8

Let $(\mathfrak{g}, [\cdot,\cdot])$ be a Lie algebra and let $r \in \mathfrak{g} \otimes \mathfrak{g}$. Then r is an r-matrix if and only if the following conditions are satisfied:

1. $\delta r_+ = 0$.
2. $\delta [\![r_-, r_-]\!] = 0$.

In this case we have a coboundary Lie bialgebra $(\mathfrak{g}, [\cdot,\cdot], [\cdot,\cdot]_r)$ with

$$[X^*, Y^*]_r = \mathrm{ad}_{\varrho_-(X^*)}^* Y^* - \mathrm{ad}_{\varrho_-(Y^*)}^* X^*,$$

for all X^*, Y^*.

► By using an explicit tensor notation it is possible to give transparent sufficient conditions for $\delta [\![r_-, r_-]\!] = 0$ to hold.

- Write $r \in \mathfrak{g} \otimes \mathfrak{g}$ as
$$r = \sum_i X_i \otimes Y_i, \qquad X_i, Y_i \in \mathfrak{g},$$
and define the following objects in the third tensor power of the enveloping algebra of $\mathfrak{g}$ (an associative algebra with unity such that $[X, Y] \in \mathfrak{g}$):
$$\begin{aligned}
r_{12} &:= \sum_i X_i \otimes Y_i \otimes 1, \\
r_{23} &:= \sum_i 1 \otimes X_i \otimes Y_i, \\
r_{13} &:= \sum_i X_i \otimes 1 \otimes Y_i.
\end{aligned}$$

- Then one has

$$[r_{12}, r_{13}] = \sum_{i,j} [X_i, X_j] \otimes Y_i \otimes Y_j,$$

$$[r_{12}, r_{23}] = \sum_{i,j} X_i \otimes [Y_i, X_j] \otimes Y_j,$$

$$[r_{13}, r_{23}] = \sum_{i,j} X_i \otimes X_j \otimes [Y_i, Y_j].$$

- It can be proved that a sufficient condition for $\delta [\![r_-, r_-]\!] = 0$ to hold is given by

$$[r_{12}, r_{13}] + [r_{12}, r_{23}] + [r_{13}, r_{23}] = 0, \tag{4.41}$$

  which is called *(classical) Yang-Baxter equation.*

- If $\mathfrak{g}$ is a loop algebra then r depends on two spectral parameters $\lambda, \mu \in \mathbb{C}$, so that $r_{ij} = r_{ij}(\lambda, \mu)$, $\lambda, \mu \in \mathbb{C}$. In many applications it happens that r_{ij} depends on the difference of the spectral parameters, i.e., $r_{ij} = r_{ij}(\lambda - \mu)$ and $r_{12}(\lambda - \mu) = -r_{21}(\mu - \lambda)$. Here $r_{21} = \Pi\, r_{12}\, \Pi$, Π being the permutation operator $\Pi(X \otimes Y) = Y \otimes X$, $X, Y \in \mathfrak{g}$. In such a case equation (4.41) takes the form

$$[r_{12}(\lambda - \mu), r_{13}(\lambda) + r_{23}(\mu)] + [r_{13}(\lambda), r_{23}(\mu)] = 0. \tag{4.42}$$

  Here $r_{12} = r \otimes 1$, $r_{23} = 1 \otimes r$ and $r_{13} = (1 \otimes \Pi) r_{12} (\Pi \otimes 1)$.

► Suppose that $r \in \mathfrak{g} \otimes \mathfrak{g}$ is an r-matrix for $(\mathfrak{g}, [\cdot,\cdot])$. Then we know that the bracket $[\cdot,\cdot]_r$ defined by

$$\begin{aligned} \langle [X^*, Y^*]_r, X \rangle &:= \langle \delta r(X), X^* \wedge Y^* \rangle \\ &= \langle [X \otimes 1 + 1 \otimes X, r], X^* \wedge Y^* \rangle \end{aligned}$$

defines a Lie algebra structure on $\mathfrak{g}^*$ for all $X^*, Y^* \in \mathfrak{g}^*$, $X \in \mathfrak{g}$. Therefore $\mathfrak{g}$ inherits a linear Poisson structure $\{\cdot,\cdot\}_r$ from $[\cdot,\cdot]_r$. We have the following definition.

Definition 4.5

Let $(\mathfrak{g}, [\cdot,\cdot])$ be a Lie algebra and let $r \in \mathfrak{g}^{\wedge 2}$ be an r-matrix. For any $F, G \in \mathscr{F}(\mathfrak{g})$ the Lie-Poisson bracket on $\mathfrak{g}$ induced by $[\cdot,\cdot]_r$,

$$\begin{aligned} \{F, G\}_r(X) &:= \langle X, [dF(X), dG(X)]_r \rangle \\ &= \langle [X \otimes 1 + 1 \otimes X, r], dF(X) \wedge dG(X) \rangle \end{aligned} \tag{4.43}$$

is called **linear *r*-bracket** *on $\mathfrak{g}$. We denote by $(\mathfrak{g}, \{\cdot,\cdot\}_r)$ the corresponding Lie-Poisson manifold.*

► Note that in (4.43) $\mathrm{d}F(\mathsf{X})$ and $\mathrm{d}G(\mathsf{X})$ are interpreted as elements of $\mathfrak{g}^*$. In particular, if we take for F and G elements X^* and Y^* of $\mathfrak{g}^*$ formula (4.43) becomes

$$\{\mathsf{X}^*,\mathsf{Y}^*\}_r(\mathsf{X}) = \langle\, [\mathsf{X}\otimes 1 + 1\otimes \mathsf{X}, r\,], \mathsf{X}^*\wedge \mathsf{Y}^*\,\rangle\,. \tag{4.44}$$

► As anticipated R-matrices and r-matrices on a finite-dimensional quadratic Lie algebra $\mathfrak{g}$ are related. It is anyway important to stress the fact that double Lie algebras and Lie bialgebras are different objects and they lead to different notions of r-matrices.

- In the case of double Lie algebras $(\mathfrak{g},[\,\cdot,\cdot\,],[\,\cdot,\cdot\,]_R)$ we are dealing with two structures of a Lie algebra on the same linear space and the associated R-matrix is a linear operator on $\mathfrak{g}$.
- By contrast, in the case of Lie bialgebras $(\mathfrak{g},[\,\cdot,\cdot\,],[\,\cdot,\cdot\,]_r)$ the brackets are defined on dual linear spaces $\mathfrak{g}$ and $\mathfrak{g}^*$ respectively and $r\in\mathfrak{g}\otimes\mathfrak{g}$.

► Let us see more concretely how R-matrices and r-matrices are related.

- Recall that there exists an isomorphism $\sigma:\mathfrak{g}\longrightarrow\mathfrak{g}^*$ which assigns to $\mathsf{X}\in\mathfrak{g}$ the linear form $\sigma(\mathsf{X})\in\mathfrak{g}^*$ according to

  $$\langle\,\sigma(\mathsf{X}),\mathsf{Y}\,\rangle = \langle\,\mathsf{X}\,|\,\mathsf{Y}\,\rangle \qquad \forall\,\mathsf{X},\mathsf{Y}\in\mathfrak{g}.$$

  The map σ allows us to identify $\mathfrak{g}$ with $\mathfrak{g}^*$.
- This leads also to an isomorphism

  $$\sigma\otimes\mathsf{Id}_{\mathfrak{g}}:\mathfrak{g}\otimes\mathfrak{g}\longrightarrow\mathfrak{g}^*\otimes\mathfrak{g}\simeq\mathsf{End}(\mathfrak{g}),$$

  which allows us to associate with every element of $\mathfrak{g}\otimes\mathfrak{g}$ a linear map $\mathfrak{g}\longrightarrow\mathfrak{g}$.

► One has the following claim.

Theorem 4.9

Let $\mathfrak{g}$ be a quadratic Lie algebra. Let $r\in\mathfrak{g}\otimes\mathfrak{g}$ and $R\in\mathsf{End}(\mathfrak{g})$ such that

$$R = 4\,(\sigma\otimes\mathsf{Id}_{\mathfrak{g}})\,r.$$

Then R is a skew-symmetric R-matrix for $\mathfrak{g}$ if and only if r is a skew-symmetric r-matrix for $\mathfrak{g}$. In this case $\sigma:(\mathfrak{g},[\,\cdot,\cdot\,],[\,\cdot,\cdot\,]_R)\longrightarrow(\mathfrak{g},[\,\cdot,\cdot\,],[\,\cdot,\cdot\,]_r)$ is a Lie algebra isomorphism.

No Proof.

4.3.3 Poisson involutivity theorems

► We now wish to apply the r-matrix formalism to integrable systems and in particular to Lax equations. Let us start with a general definition of Lax operator.

- Let $(\mathcal{M}, \{\cdot,\cdot\})$ be a Poisson manifold and $(\mathfrak{g}, [\cdot,\cdot])$ be a n-dimensional Lie algebra (even a loop algebra is admissible). We denote by $\{\mathsf{E}_1, \ldots, \mathsf{E}_n\}$ an arbitrary basis on $\mathfrak{g}$ and by $\{\mathsf{E}_1^*, \ldots, \mathsf{E}_n^*\}$ the dual basis.

- To be more explicit, we denote by $\mathscr{F}(\mathcal{M}, \mathfrak{g})$ the space of functions on $\mathcal{M}$ with values in $\mathfrak{g}$. Elements of $\mathscr{F}(\mathcal{M}, \mathfrak{g})$ may be viewed as matrices belonging to $\mathfrak{g}$ whose entries are smooth functions on $\mathcal{M}$. hence the Lie bracket $[\cdot,\cdot]$ on $\mathfrak{g}$ leads to a Lie bracket on $\mathscr{F}(\mathcal{M}, \mathfrak{g})$, which is also denoted by $[\cdot,\cdot]$.

- We equip $\mathcal{M}$ with a Poisson structure $\{\cdot,\cdot\}_\mathfrak{g}$ coming from $\mathfrak{g}$. This can be a Lie-Poisson structure, an R-bracket or an r-bracket. We denote by $(\mathfrak{g}, \{\cdot,\cdot\}_\mathfrak{g})$ the corresponding Poisson manifold.

- For $\mathsf{X}, \mathsf{Y} \in \mathscr{F}(\mathcal{M}, \mathfrak{g})$ we now define an element $\{\mathsf{X} \overset{\otimes}{,} \mathsf{Y}\}_\mathfrak{g} \in \mathscr{F}(\mathcal{M}, \mathfrak{g} \otimes \mathfrak{g})$ by setting

$$\{\mathsf{X} \overset{\otimes}{,} \mathsf{Y}\}_\mathfrak{g} := \sum_{i,j=1}^{n} \left\{\mathsf{X}^\star \mathsf{E}_i^*, \mathsf{Y}^\star \mathsf{E}_j^*\right\}_\mathfrak{g} \mathsf{E}_i \otimes \mathsf{E}_j, \tag{4.45}$$

  which is easily seen to be independent of the chosen basis for $\mathfrak{g}$.

Definition 4.6

An element $\mathsf{X} \in \mathscr{F}(\mathcal{M}, \mathfrak{g})$ *is called a* **Lax operator** *if*

$$\mathsf{X} : (\mathcal{M}, \{\cdot,\cdot\}) \to (\mathfrak{g}, \{\cdot,\cdot\}_\mathfrak{g})$$

is a Poisson map.

► We now give the following claim which contains a remarkable formula which sometimes goes under the name of *first russian formula* (the *second russian formula* refers to a quadratic r-matrix structure which is not considered in this course).

Theorem 4.10

If $\mathsf{X} \in \mathscr{F}(\mathcal{M}, \mathfrak{g})$ *is a Lax operator in* $\mathfrak{g}$ *with* r*-matrix* r *then*

$$\{\mathsf{X} \overset{\otimes}{,} \mathsf{X}\}_r = [\mathsf{X} \otimes 1 + 1 \otimes \mathsf{X}, r]. \tag{4.46}$$

Proof. If X is a Lax operator in $\mathfrak{g}$ with r-matrix r then formula (4.45) gives

$$\begin{aligned}\{\mathsf{X} \overset{\otimes}{,} \mathsf{X}\}_r &:= \sum_{i,j=1}^{n} \mathsf{X}^\star \left\{ \mathsf{E}_i^*, \mathsf{E}_j^* \right\}_r \mathsf{E}_i \otimes \mathsf{E}_j \\ &= \sum_{i,j=1}^{n} \left\{ \mathsf{E}_i^*, \mathsf{E}_j^* \right\}_r (\mathsf{X})\, \mathsf{E}_i \otimes \mathsf{E}_j.\end{aligned}$$

Using (4.44) we get

$$\begin{aligned}\{\mathsf{X} \overset{\otimes}{,} \mathsf{X}\}_r &= \sum_{i,j=1}^{n} \left\langle [\mathsf{X} \otimes 1 + 1 \otimes \mathsf{X}, r], \mathsf{E}_i^* \wedge \mathsf{E}_j^* \right\rangle \mathsf{E}_i \otimes \mathsf{E}_j \\ &= [\mathsf{X} \otimes 1 + 1 \otimes \mathsf{X}, r],\end{aligned}$$

which is the claim. ∎

► Let us observe that in Theorem 4.46 r is necessarily a skew-symmetric element of $\mathfrak{g} \otimes \mathfrak{g}$. This does not mean that if we represent r in terms of a matrix then such a matrix is skew-symmetric.

► Hereafter we assume that $\{\cdot, \cdot\}_{\mathfrak{g}}$ comes from an r-matrix and $\mathfrak{g} = \mathfrak{gl}(n, \mathbb{R})$.

- We consider a Lax operator $\mathsf{X} = (\mathsf{X}_{ij})_{1 \leqslant i,j \leqslant n}$ in $\mathfrak{g}$. Here X_{ij} is considered as a function on $\mathcal{M}$.
- The $n \times n$ matrices E_{ij}, $i, j = 1, \ldots, n$, with matrix elements $(\mathsf{E}_{ij})_{k\ell} := \delta_{ik}\,\delta_{j\ell}$ form a basis for $\mathfrak{g}$. For any element $\mathsf{X} \in \mathfrak{g}$ we have the representation

$$\mathsf{X} = \sum_{i,j=1}^{n} \mathsf{X}_{ij}\, \mathsf{E}_{ij}.$$

- We construct the following $n^2 \times n^2$ matrices belonging to $\mathscr{F}(\mathcal{M}, \mathfrak{g} \otimes \mathfrak{g})$:

$$\mathsf{X} \otimes 1 = \sum_{i,j=1}^{n} \mathsf{X}_{ij}\, \mathsf{E}_{ij} \otimes 1 = \begin{pmatrix} \mathsf{X}_{11} & 0 & \cdots & 0 & \cdots & \mathsf{X}_{1n} & 0 & \cdots & 0 \\ 0 & \mathsf{X}_{11} & \cdots & 0 & \cdots & 0 & \mathsf{X}_{1n} & \cdots & 0 \\ \vdots & \vdots & \ddots & \vdots & \vdots & \vdots & \vdots & \ddots & \vdots \\ 0 & 0 & \cdots & \mathsf{X}_{11} & \cdots & 0 & 0 & \cdots & \mathsf{X}_{1n} \\ \vdots & \vdots & \vdots & \vdots & \vdots & \vdots & \vdots & \vdots & \vdots \\ \mathsf{X}_{n1} & 0 & \cdots & 0 & \cdots & \mathsf{X}_{nn} & 0 & \cdots & 0 \\ 0 & \mathsf{X}_{n1} & \cdots & 0 & \cdots & 0 & \mathsf{X}_{nn} & \cdots & 0 \\ \vdots & \vdots & \ddots & \vdots & \vdots & \vdots & \vdots & \ddots & \vdots \\ 0 & 0 & \cdots & \mathsf{X}_{n1} & \cdots & 0 & 0 & \cdots & \mathsf{X}_{nn} \end{pmatrix},$$

and

$$1 \otimes \mathsf{X} = \sum_{i,j=1}^{n} \mathsf{X}_{ij} \, 1 \otimes \mathsf{E}_{ij} = \begin{pmatrix} \mathsf{X} & 0 & \cdots & 0 \\ 0 & \mathsf{X} & \cdots & 0 \\ \vdots & \vdots & \ddots & \vdots \\ 0 & 0 & \cdots & \mathsf{X} \end{pmatrix}.$$

- Then $\left\{\mathsf{X} \overset{\otimes}{,} \mathsf{X}\right\}_r \in \mathscr{F}(\mathcal{M}, \mathfrak{g} \otimes \mathfrak{g})$ is the $n^2 \times n^2$ matrix which encodes all possible Poisson brackets between the functions X_{ij}:

$$\{\mathsf{X} \overset{\otimes}{,} \mathsf{X}\}_r = \sum_{i,j,k,\ell=1}^{n} \left\{ \mathsf{X}_{ij}, \mathsf{X}_{k\ell} \right\}_r \mathsf{E}_{ij} \otimes \mathsf{E}_{k\ell}.$$

Explicitly,

$$\{\mathsf{X} \overset{\otimes}{,} \mathsf{X}\}_r = \begin{pmatrix} \{\mathsf{X}_{11}, \mathsf{X}_{11}\}_r & \cdots & \{\mathsf{X}_{11}, \mathsf{X}_{1n}\}_r & \cdots & \{\mathsf{X}_{1n}, \mathsf{X}_{11}\}_r & \cdots & \{\mathsf{X}_{1n}, \mathsf{X}_{1n}\}_r \\ \vdots & & \vdots & & \vdots & & \vdots \\ \{\mathsf{X}_{11}, \mathsf{X}_{n1}\}_r & \cdots & \{\mathsf{X}_{11}, \mathsf{X}_{nn}\}_r & \cdots & \{\mathsf{X}_{1n}, \mathsf{X}_{n1}\}_r & \cdots & \{\mathsf{X}_{1n}, \mathsf{X}_{nn}\}_r \\ \vdots & \vdots & \vdots & \vdots & \vdots & \vdots & \vdots \\ \{\mathsf{X}_{n1}, \mathsf{X}_{11}\}_r & \cdots & \{\mathsf{X}_{n1}, \mathsf{X}_{1n}\}_r & \cdots & \{\mathsf{X}_{nn}, \mathsf{X}_{11}\}_r & \cdots & \{\mathsf{X}_{nn}, \mathsf{X}_{1n}\}_r \\ \vdots & & \vdots & & \vdots & & \vdots \\ \{\mathsf{X}_{n1}, \mathsf{X}_{n1}\}_r & \cdots & \{\mathsf{X}_{n1}, \mathsf{X}_{nn}\}_r & \cdots & \{\mathsf{X}_{nn}, \mathsf{X}_{n1}\}_r & \cdots & \{\mathsf{X}_{nn}, \mathsf{X}_{nn}\}_r \end{pmatrix}.$$

► Let us give some useful formulas for $\{ \cdot \overset{\otimes}{,} \cdot \}_r$ (without proof):

- For $\mathsf{Y}, \mathsf{X} \in \mathscr{F}(\mathcal{M}, \mathfrak{g})$ there holds

$$\mathsf{Trace}(\{\mathsf{X} \overset{\otimes}{,} \mathsf{Y}\}_r) = \{\mathsf{Trace}\,\mathsf{X}, \mathsf{Trace}\,\mathsf{Y}\}_r. \tag{4.47}$$

- Let $\mathsf{X} \in \mathscr{F}(\mathcal{M}, \mathfrak{g})$ be a Lax operator in $\mathfrak{g}$ with r-matrix r. For $k, \ell \in \mathbb{N}$ there holds

$$\left\{\mathsf{X}^k \overset{\otimes}{,} \mathsf{X}^\ell\right\}_r = \sum_{i=0}^{k-1} \sum_{j=0}^{\ell-1} \left[\left(\mathsf{X}^{k-i-1} \otimes \mathsf{X}^{\ell-j-1}\right) r \left(\mathsf{X}^i \otimes \mathsf{X}^j\right), \mathsf{X} \otimes 1 + 1 \otimes \mathsf{X} \right]. \tag{4.48}$$

This formula generalizes (4.46). Notice that each term in the r.h.s. is a commutator. Therefore

$$\mathsf{Trace}\left(\left\{\mathsf{X}^k \overset{\otimes}{,} \mathsf{X}^\ell\right\}_r\right) = 0 \qquad \forall\, k, \ell \in \mathbb{N}.$$

Using (4.47) there follows that

$$\left\{\mathsf{Trace}\,\mathsf{X}^k, \mathsf{Trace}\,\mathsf{X}^\ell\right\}_r = 0 \qquad \forall\, k, \ell \in \mathbb{N}. \tag{4.49}$$

- Let $\mathsf{X}, \mathsf{Y} \in \mathscr{F}(\mathcal{M}, \mathfrak{g})$ and consider $\mathsf{X} \otimes \mathsf{Y} \in \mathscr{F}(\mathcal{M}, \mathfrak{g} \otimes \mathfrak{g})$. We define the traces $\mathsf{Trace}_i : \mathfrak{g} \otimes \mathfrak{g} \to \mathfrak{g}$, $i = 1, 2$, by setting

$$\mathsf{Trace}_1(\mathsf{X} \otimes \mathsf{Y}) := (\mathsf{Trace}\,\mathsf{X})\,\mathsf{Y}, \qquad \mathsf{Trace}_2(\mathsf{X} \otimes \mathsf{Y}) := (\mathsf{Trace}\,\mathsf{Y})\,\mathsf{X}.$$

► The role of r-matrices for the theory of integrable systems is made explicit in the following statement.

Theorem 4.11

Let $\mathsf{X} \in \mathscr{F}(\mathcal{M}, \mathfrak{g})$ be a Lax operator in $\mathfrak{g}$ with r-matrix r.

1. *The functions $H_k \in \mathscr{F}(\mathcal{M})$ defined by*

$$H_k := \mathsf{Trace}\,\mathsf{X}^k, \qquad k \in \mathbb{N}, \tag{4.50}$$

are in Poisson involution w.r.t. $\{\cdot,\cdot\}_r$ (and w.r.t. $\{\cdot,\cdot\}$).

2. *The Hamiltonian vector fields associated with H_k have the Lax form*

$$\mathsf{v}_{H_k}(\mathsf{X}) = \left[k\,\mathsf{Trace}_2\left(1 \otimes \mathsf{X}^{k-1} r\right), \mathsf{X}\right], \qquad k \in \mathbb{N}.$$

Proof. The first claim is a consequence of (4.49). Let us prove the second claim. $\mathsf{v}_{H_i}(\mathsf{X})$ is a matrix whose entry at position (i, ℓ) is given by

$$\mathsf{v}_{H_k}(\mathsf{X}_{i\ell}) = \left\{\mathsf{X}_{i\ell}, \mathsf{Trace}\,\mathsf{X}^k\right\}_r = \left(\mathsf{Trace}_2\left\{\mathsf{X} \overset{\otimes}{,} \mathsf{X}^k\right\}_r\right)_{i\ell}.$$

Using (4.48) it follows that

$$\begin{aligned} \mathsf{v}_{H_k}(\mathsf{X}) &= \mathsf{Trace}_2\left\{\mathsf{X} \overset{\otimes}{,} \mathsf{X}^k\right\}_r \\ &= \sum_{j=0}^{k-1} \mathsf{Trace}_2\left[\left(1 \otimes \mathsf{X}^{k-j-1}\right) r \left(1 \otimes \mathsf{X}^j\right), \mathsf{X} \otimes 1 + 1 \otimes \mathsf{X}\right] \\ &= \left[k\,\mathsf{Trace}_2\left(1 \otimes \mathsf{X}^{k-1} r\right), \mathsf{X}\right]. \end{aligned}$$

In order to prove the last equality, one may first assume that r is a decomposable tensor and then write the bracket as a commutator. The proof for general r then follows from the fact that any r is a finite sum of decomposable tensors. ∎

► Generalizing the proof of Theorem 4.11 one can prove the following claim.

Theorem 4.12

Let $\mathsf{X} \in \mathscr{F}(\mathcal{M},\mathfrak{g})$ be a Lax operator in $\mathfrak{g}$ with r-matrix r. If there exist functions $r_1, r_2 \in \mathscr{F}(\mathcal{M},\mathfrak{g}\otimes\mathfrak{g})$ such that

$$\{\mathsf{X} \overset{\otimes}{,} \mathsf{X}\}_{r_1,r_2} = [\,1\otimes\mathsf{X}, r_1\,] + [\,\mathsf{X}\otimes 1, r_2\,], \tag{4.51}$$

then the functions defined by (4.50) are in Poisson involution w.r.t. $\{\cdot,\cdot\}$.

► Remarkably, one can show that, conversely, if the traces of X are in Poisson involution then - under some conditions on X - there exist functions r_1, r_2, defined on a dense open subset of $\mathcal{M}$, such that (4.51) holds. This is contained in the following remarkable claim.

Theorem 4.13 (*Babelon-Viallet*)

Let $\mathsf{X} \in \mathscr{F}(\mathcal{M},\mathfrak{g})$ be a Lax operator in $\mathfrak{g}$ with r-matrix r. Assume that:

1. *There exists an open dense subset $\mathcal{U}$ of $\mathcal{M}$ such that X is diagonalizable in $\mathcal{U}$.*
2. *The coefficients of the characteristic polynomial of X are in Poisson involution w.r.t. $\{\cdot,\cdot\}$.*

Then there exists an open dense subset $\mathcal{V} \subseteq \mathcal{U}$ of $\mathcal{M}$, and smooth functions $r_1, r_2 : \mathcal{V} \to \mathfrak{g}\otimes\mathfrak{g}$ such that equation (4.51) holds true.

No Proof.

4.3.4 *Tensor formulation of the linear R-bracket on $\mathfrak{g}$*

► We now discuss the tensor formulation of the linear R-bracket on $\mathfrak{g}$:

$$\begin{aligned} \{F,G\}_R(\mathsf{X}) \;&:=\; \frac{1}{2}\langle \mathsf{X}\,|\,[\,R(\nabla F(\mathsf{X})), \nabla G(\mathsf{X})\,]\rangle \\ &+\; \frac{1}{2}\langle \mathsf{X}\,|\,[\,\nabla F(\mathsf{X}), R(\nabla G(\mathsf{X}))\,]\rangle, \end{aligned} \tag{4.52}$$

where $F, G \in \mathscr{F}(\mathfrak{g})$ and $R \in \mathsf{End}(\mathfrak{g})$. We will end up with the r-matrix formulation of (4.52) in the same spirit of Theorem 4.9.

► We proceed as in Subsection 2.5.4.

- Let $\{\mathsf{E}_1,\ldots,\mathsf{E}_n\}$ be an arbitrary basis on $\mathfrak{g}$. Denoting by c_{ij}^k the structure constants of $\mathfrak{g}$ in the basis $\{\mathsf{E}_1,\ldots,\mathsf{E}_n\}$, we have

$$[\,\mathsf{E}_i, \mathsf{E}_j\,] = \sum_{k=1}^{n} c_{ij}^k\,\mathsf{E}_k.$$

- An element of $\mathfrak{g}$ can be represented as

$$\mathsf{X} = \sum_{i=1}^{n} \xi_i \, \mathsf{E}_i, \tag{4.53}$$

for some smooth coordinates ξ_i which form a functional basis of $\mathscr{F}(\mathfrak{g})$.

- Let $\{\mathsf{E}_1^*, \ldots, \mathsf{E}_n^*\}$ be the dual basis of $\mathfrak{g}^*$ with $\langle \mathsf{E}_i, \mathsf{E}_j^* \rangle = \delta_{ij}$. We have:

$$\left\langle \mathsf{X}, \mathsf{E}_j^* \right\rangle = \sum_{i=1}^{n} \xi_i \left\langle \mathsf{E}_i, \mathsf{E}_j^* \right\rangle = \xi_j,$$

from which there follows that $\nabla \xi_j(\mathsf{X}) = \mathsf{E}_j^*$.

- If A is the inverse of the $n \times n$ invertible matrix A^{-1} with entries $\langle \mathsf{E}_i \,|\, \mathsf{E}_j \rangle$, we can write

$$\mathsf{E}_i^* = \sum_{k=1}^{n} A_{ik} \, \mathsf{E}_k, \qquad A_{ik} = \langle \mathsf{E}_i^* \,|\, \mathsf{E}_k^* \rangle. \tag{4.54}$$

This formula encodes the identification of $\mathfrak{g}$ with $\mathfrak{g}^*$ by means of $\langle \cdot \,|\, \cdot \rangle$. Therefore if we write a Lie bracket $[\, \mathsf{E}_i^*, \mathsf{E}_j^* \,]$ between elements of $\mathfrak{g}^*$ this has to be interpreted according to (4.54). Note also that

$$\langle \mathsf{E}_i^* \,|\, \mathsf{E}_j \rangle = \sum_{k=1}^{n} A_{ik} \langle \mathsf{E}_k \,|\, \mathsf{E}_j \rangle = \sum_{k=1}^{n} \langle \mathsf{E}_i^* \,|\, \mathsf{E}_k^* \rangle \langle \mathsf{E}_k \,|\, \mathsf{E}_j \rangle = \delta_{ij}.$$

- We now define the endomorphism of $\mathfrak{g}$ given by

$$R(\mathsf{E}_i^*) := \sum_{k=1}^{n} r_{ik} \mathsf{E}_k, \qquad r_{ik} = \langle \, R(\mathsf{E}_i^*), \mathsf{E}_k^* \, \rangle \, . \tag{4.55}$$

Theorem 4.14

The coordinate representation of the linear R-bracket on $\mathfrak{g}$ takes the following form:

$$\{ \, \xi_i, \xi_j \, \}_R(\mathsf{X}) = \frac{1}{2} \sum_{k,\ell=1}^{n} \xi_\ell \left(r_{ik} \, c_{\ell k}^{j} - r_{jk} \, c_{\ell k}^{i} \right) . \tag{4.56}$$

Proof. We have:

$$\begin{aligned}
2\,\{\xi_i, \xi_j\}_R \;&:=\; \langle \mathsf{X} \,|\; [\, R(\nabla \xi_i(\mathsf{X})), \nabla \xi_j(\mathsf{X}) \,] \,\rangle + \langle \mathsf{X} \,|\; [\, \nabla \xi_i(\mathsf{X}), R(\nabla \xi_j(\mathsf{X})) \,] \,\rangle \\
&=\; \left\langle \mathsf{X} \middle| \left[R(\mathsf{E}_i^*), \mathsf{E}_j^* \right] \right\rangle + \left\langle \mathsf{X} \middle| \left[\mathsf{E}_i^*, R(\mathsf{E}_j^*) \right] \right\rangle \\
&=\; \left\langle \mathsf{E}_j^* \middle| \, [\mathsf{X}, R(\mathsf{E}_i^*)] \right\rangle - \left\langle \mathsf{E}_i^* \middle| \, [\mathsf{X}, R(\mathsf{E}_j^*)] \right\rangle ,
\end{aligned}$$

where we used (1.49). Now, using (4.53), (4.54) and (4.55) we get

$$
\begin{aligned}
2\{\xi_i,\xi_j\}_R &= \sum_{k,\ell=1}^{n} \xi_\ell\, r_{ik}\, \langle \mathsf{E}_j^* \mid [\mathsf{E}_\ell,\mathsf{E}_k]\rangle - \sum_{k,\ell=1}^{n} \xi_\ell\, r_{jk}\, \langle \mathsf{E}_i^* \mid [\mathsf{E}_\ell,\mathsf{E}_k]\rangle \\
&= \sum_{k,\ell,s=1}^{n} \xi_\ell\, r_{ik}\, c_{\ell k}^{s}\, \langle \mathsf{E}_j^* \mid \mathsf{E}_s\rangle - \sum_{k,\ell,s=1}^{n} \xi_\ell\, r_{jk}\, c_{\ell k}^{s}\, \langle \mathsf{E}_i^* \mid \mathsf{E}_s\rangle \\
&= \sum_{k,\ell=1}^{n} \xi_\ell\, r_{ik}\, c_{\ell k}^{j} - \sum_{k,\ell=1}^{d} \xi_\ell\, r_{jk}\, c_{\ell k}^{i},
\end{aligned}
$$

which is the desired formula. ■

► To put (4.56) in a compact form we introduce the r-matrix $r \in \mathfrak{g}\otimes\mathfrak{g}$ canonically corresponding to $R \in \mathsf{End}(\mathfrak{g})$ by setting

$$
r := \frac{1}{2}\sum_{i,k=1}^{n} r_{ik}\, \mathsf{E}_i \otimes \mathsf{E}_k,
$$

We also define

$$
r^* := \frac{1}{2}\sum_{i,k=1}^{n} r_{ki}\, \mathsf{E}_i \otimes \mathsf{E}_k.
$$

► Now we have the following claim.

Theorem 4.15

The linear R-bracket on $\mathfrak{g}$ admits the following tensor formulation:

$$
\{\mathsf{X} \overset{\otimes}{,} \mathsf{X}\}_R = [1\otimes\mathsf{X}, r] - [\mathsf{X}\otimes 1, r^*]. \tag{4.57}
$$

Proof. We have:

$$
\begin{aligned}
[1\otimes\mathsf{X}, r] &= \frac{1}{2}\left[1\otimes\sum_{\ell=1}^{n}\xi_\ell\,\mathsf{E}_\ell,\ \sum_{i,k=1}^{n} r_{ik}\,\mathsf{E}_i\otimes\mathsf{E}_k\right] \\
&= \frac{1}{2}\sum_{i,k,\ell=1}^{n} \xi_\ell\, r_{ik}\, \mathsf{E}_i \otimes [\mathsf{E}_\ell,\mathsf{E}_k] \\
&= \frac{1}{2}\sum_{i,j,k,\ell=1}^{n} \xi_\ell\, r_{ik}\, c_{\ell k}^{j}\, \mathsf{E}_i\otimes\mathsf{E}_j.
\end{aligned} \tag{4.58}
$$

Similarly one finds:

$$
[\mathsf{X}\otimes 1, r^*] = \frac{1}{2}\sum_{i,j,k,\ell=1}^{n} \xi_\ell\, r_{jk}\, c_{\ell k}^{i}\, \mathsf{E}_i\otimes\mathsf{E}_j. \tag{4.59}
$$

From (4.58), (4.59) and (4.56) we get the desired formula. ■

5

Examples: Toda, Garnier, Gaudin, Lagrange

5.1 The open-end Toda lattice

► One of the most famous integrable lattice systems is undoubtedly the Toda lattice (1967). We will consider only the so called *open-end Toda lattice*, namely a Toda lattice defined on a one-dimensional lattice with open-end boundary conditions.

5.1.1 *Definition of the phase space and equations of motion*

► The Newton equations of motion of the open-end Toda lattice read:

$$\ddot{q}_i = \mathrm{e}^{q_{i+1}-q_i} - \mathrm{e}^{q_i - q_{i-1}}, \qquad i = 1, \dots, N, \tag{5.1}$$

with $q_0 = \infty$, $q_{N+1} = -\infty$. They describe the dynamics of N ideal particles distributed on a one-dimensional lattice that interact with nearest neighbors via forces exponentially depending on distances.

- Equations (5.1) define a canonical Hamiltonian system in the canonical space $\mathbb{R}^{2N}$ with coordinates $(q,p) := (q_1, \dots, q_N, p_1, \dots, p_N)$. The Hamiltonian is

$$H := \sum_{i=1}^{N} \left(\frac{p_i^2}{2} + \mathrm{e}^{q_{i+1}-q_i} \right), \tag{5.2}$$

  and the corresponding Hamilton equations are

$$\begin{cases} \dot{q}_i = p_i, \\ \dot{p}_i = \mathrm{e}^{q_{i+1}-q_i} - \mathrm{e}^{q_i - q_{i-1}}, \end{cases} \tag{5.3}$$

  with $i = 1, \dots, N$.

- We define *Flaschka variables* $(a,b) := (a_1, \dots, a_N, b_1, \dots, b_N)$:

$$a_i := \mathrm{e}^{q_{i+1}-q_i}, \qquad b_i := \dot{q}_i = p_i, \qquad i = 1, \dots, N. \tag{5.4}$$

  They allow us to transform (5.3) into a quadratic system of differential equations:

$$\begin{cases} \dot{a}_i = a_i(b_{i+1} - b_i), \\ \dot{b}_i = a_i - a_{i-1}, \end{cases} \tag{5.5}$$

  with $i = 1, \dots, N$, and boundary conditions $a_0 \equiv a_N \equiv 0$. We denote by $\mathcal{M}$ the $(2N-1)$-dimensional phase space parametrized by coordinates (a,b):

$$\mathcal{M} := \left\{ (a,b) \in \mathbb{R}^{2N} \, : \, a_N \equiv 0 \right\} \simeq \mathbb{R}^{2N-1}.$$

The space $\mathcal{M}$ is no longer a symplectic vector space. Indeed, the Poisson brackets defined on $\mathcal{M}$, which are given by the pull-back under (5.4) of the canonical Poisson brackets on $\mathbb{R}^{2N}$ are now degenerate.

Theorem 5.1

1. *The space $\mathcal{M}$ is a Poisson manifold with (non-vanishing) Poisson brackets*

$$\{ a_i, b_i \} = -a_i, \qquad \{ a_i, b_{i+1} \} = a_i. \tag{5.6}$$

The Poisson structure (5.6) has rank $2(N-1)$. It has only one Casimir given by

$$C := \sum_{i=1}^{N} b_i. \tag{5.7}$$

2. *Equations (5.5) are Hamiltonian w.r.t. the Poisson structure (5.6) with Hamiltonian*

$$H := \sum_{i=1}^{N} \left(\frac{b_i^2}{2} + a_i \right). \tag{5.8}$$

Proof. Both claims can be easily proved by direct computation.

1. We compute:

$$\{ a_i, b_i \} = \sum_{j=1}^{N} \left(\frac{\partial}{\partial q_j} e^{q_{i+1}-q_i} \frac{\partial}{\partial p_j} p_i - \frac{\partial}{\partial p_j} e^{q_{i+1}-q_i} \frac{\partial}{\partial q_j} p_i \right) = -a_i.$$

Similarly,

$$\{ a_i, b_{i+1} \} = \sum_{j=1}^{N} \left(\frac{\partial}{\partial q_j} e^{q_{i+1}-q_i} \frac{\partial}{\partial p_j} p_{i+1} - \frac{\partial}{\partial p_j} e^{q_{i+1}-q_i} \frac{\partial}{\partial q_j} p_{i+1} \right) = a_i.$$

All other Poisson brackets vanish. One can easily check that the rank of the Poisson matrix defined in terms of (5.6) is $2(N-1)$. The function C defined in (5.7) is in Poisson involution with any other function defined on $\mathcal{M}$.

2. The Hamiltonian (5.8) is simply the pull-back of the canonical Hamiltonian (5.2) under (5.4). We compute:

$$\dot{a}_i = \{ a_i, H \} = \frac{1}{2} \left(\left\{ a_i, b_i^2 \right\} + \left\{ a_i, b_{i+1}^2 \right\} \right) = a_i(b_{i+1} - b_i).$$

Similarly for $\dot{b}_i$:

$$\dot{b}_i = \{ b_i, H \} = \{ b_i, a_i \} + \{ b_i, a_{i-1} \} = a_i - a_{i-1}.$$

The Theorem is proved. ■

► One can show that the open-end Toda lattice admits a tri-Hamiltonian structure, i.e., the system is Hamiltonian w.r.t. three independent and compatible Poisson structures.

5.1.2 Algebraic setting

► We now introduce the basic Lie algebras which allow us to construct a Lax representation for equations (5.5).

- The basic Lie algebra is $\mathfrak{g} = \mathfrak{gl}(N, \mathbb{R})$. As a basis we consider the $N \times N$ matrices E_{ij}, $i, j = 1, \dots, N$, so that

$$[\, \mathsf{E}_{ij}, \mathsf{E}_{k\ell}\,] = \delta_{jk}\, \mathsf{E}_{i\ell} - \delta_{\ell i}\, \mathsf{E}_{kj}.$$

  We choose the Ad-invariant bilinear symmetric non-degenerate form $\langle \mathsf{X} \,|\, \mathsf{Y} \rangle := \mathsf{Trace}(\mathsf{X}\,\mathsf{Y})$. In particular, $\langle \mathsf{E}_{mn} \,|\, \mathsf{E}_{jk} \rangle = \delta_{mk}\delta_{nj}$.

- As a linear space $\mathfrak{g}$ can be decomposed as

$$\mathfrak{g} = \bigoplus_{i=-N+1}^{N-1} \mathfrak{g}_i,$$

  where $\mathfrak{g}_i$ is the set of matrices with non-zero entries only on the i-th diagonal, i.e., in the positions (j,k) with $j - k = i$. In particular, $\mathfrak{g}_0$ is a set of diagonal matrices, which serves as a commutative subalgebra of $\mathfrak{g}$.

- We introduce the subalgebras of $\mathfrak{g}$:

$$\mathfrak{g}_+ := \bigoplus_{i=0}^{N-1} \mathfrak{g}_i, \qquad \mathfrak{g}_- := \bigoplus_{i=-N+1}^{-1} \mathfrak{g}_i,$$

  so that $\mathfrak{g} = \mathfrak{g}_+ \oplus \mathfrak{g}_-$. Accordingly, we denote by $\pi_\pm : \mathfrak{g} \to \mathfrak{g}_\pm$ the corresponding projection operators. We also define the linear operator $R := \pi_+ - \pi_-$.

5.1.3 Complete integrability

► The complete integrability of the vector field (5.5) is established once we have $2N - 1 - (N - 1) = N$ integrals of motion which are Poisson involutive and independent.

► In the next claim we give a Lax formulation for equations of motion (5.5).

Theorem 5.2

1. *Equations (5.5) admit the following Lax representation on the Lie algebra* $\mathfrak{gl}(N,\mathbb{R})$*:*

$$\dot{\mathsf{X}} = \frac{1}{2}[\mathsf{X}, R(\mathsf{X})] = [\mathsf{X}, \mathsf{B}_+] = -[\mathsf{X}, \mathsf{B}_-], \tag{5.9}$$

where the Lax operator is given by

$$\mathsf{X} := \sum_{i=1}^{N} (a_i\, \mathsf{E}_{i,i+1} + b_i\, \mathsf{E}_{ii} + \mathsf{E}_{i+1,i}), \tag{5.10}$$

while the auxiliary matrices are

$$\mathsf{B}_+ := \pi_+(\mathsf{X}) = \sum_{i=1}^{N} (b_i\, \mathsf{E}_{ii} + \mathsf{E}_{i+1,i}),$$

$$\mathsf{B}_- := \pi_-(\mathsf{X}) = \sum_{i=1}^{N} a_i\, \mathsf{E}_{i,i+1}.$$

Here we set $\mathsf{E}_{N+1,N} \equiv \mathsf{E}_{N,N+1} \equiv 0$.

2. *The spectral invariants of the Lax operator (5.10) provide a set of* N *independent integrals of motion of (5.5). In particular,*

$$C = \operatorname{Trace} \mathsf{X}, \qquad H = \frac{1}{2} \operatorname{Trace} \mathsf{X}^2.$$

Proof. The Theorem is indeed a consequence of the general theory (see Section 4.2). There remain only two points to clarify. The first one is to show the equivalence of equations of motion (5.5) and their Lax representation (5.9). The second one is to prove that the spectral invariants of X actually provide N independent integrals of motion. We shall prove only the first point, by illustrating the computation in the simplest case $\dot{\mathsf{X}} = -[\mathsf{X}, \mathsf{B}_-]$. In this case we have:

$$\mathsf{X} = \begin{pmatrix} b_1 & a_1 & 0 & \dots & 0 & 0 \\ 1 & b_2 & a_2 & \dots & 0 & 0 \\ 0 & 1 & b_3 & \dots & 0 & 0 \\ \vdots & \vdots & \vdots & \ddots & \vdots & \vdots \\ 0 & 0 & 0 & \dots & b_{N-1} & a_{N-1} \\ 0 & 0 & 0 & \dots & 1 & b_N \end{pmatrix}, \quad \mathsf{B}_- = \begin{pmatrix} 0 & a_1 & 0 & \dots & 0 & 0 \\ 0 & 0 & a_2 & \dots & 0 & 0 \\ 0 & 0 & 0 & \dots & 0 & 0 \\ \vdots & \vdots & \vdots & \ddots & \vdots & \vdots \\ 0 & 0 & 0 & \dots & 0 & a_{N-1} \\ 0 & 0 & 0 & \dots & 0 & 0 \end{pmatrix}.$$

Now, a direct verification shows the claim. ∎

Example 5.1 (*The case* $N = 3$)

Let us fix $N = 3$.

- The only non-vanishing Poisson brackets are

$$\{a_1, b_1\} = -a_1, \qquad \{a_2, b_2\} = -a_1, \qquad \{a_1, b_2\} = a_1, \qquad \{a_2, b_3\} = a_2.$$

  The corresponding 6×6 Poisson matrix has rank 4 and it admits $C := b_1 + b_2 + b_3$ as Casimir.

- The Hamiltonian reads

$$H := \frac{1}{2}\left(b_1^2 + b_2^2 + b_3^2\right) + a_1 + a_2.$$

- The Lax operator is

$$\mathsf{X} := \begin{pmatrix} b_1 & a_1 & 0 \\ 1 & b_2 & a_2 \\ 0 & 1 & b_3 \end{pmatrix},$$

  which gives 3 independent spectral invariants: $C = \operatorname{Trace}\mathsf{X}$, $H = \operatorname{Trace}\mathsf{X}^2/2$, and

$$F := \frac{1}{3}\operatorname{Trace}\mathsf{X}^3 = \frac{1}{3}\left(b_1^3 + b_2^3 + b_3^3\right) + a_1(b_1 + b_2) + a_2(b_2 + b_3).$$

► The fact that the spectral invariants of the Lax operator (5.9) are Poisson involutive on $\mathcal{M}$ is a consequence of the next Theorem combined with Theorem 4.12.

Theorem 5.3

The Lax operator (5.10) satisfies a linear r*-matrix bracket*

$$\{\mathsf{X} \overset{\otimes}{,} \mathsf{X}\} = [\mathsf{X} \otimes 1, r_0 - r] - [1 \otimes \mathsf{X}, r + r_0], \tag{5.11}$$

where

$$r := \frac{1}{2}\left(\sum_{i<j} \mathsf{E}_{ij} \otimes \mathsf{E}_{ji} - \sum_{i>j} \mathsf{E}_{ij} \otimes \mathsf{E}_{ji}\right),$$

and

$$r_0 := \frac{1}{2}\sum_{i=1}^{N} \mathsf{E}_{ii} \otimes \mathsf{E}_{ii}.$$

Proof. Instead of proving the general claim we verify (5.11) for the case $N = 3$.

- A direct computation shows that

$$\{\mathsf{X} \overset{\otimes}{,} \mathsf{X}\} = \begin{pmatrix} 0 & a_1 & 0 & -a_1 & 0 & 0 & 0 & 0 & 0 \\ 0 & 0 & 0 & 0 & a_1 & 0 & 0 & 0 & 0 \\ 0 & 0 & 0 & 0 & 0 & 0 & 0 & 0 & 0 \\ 0 & 0 & 0 & 0 & -a_1 & 0 & 0 & 0 & 0 \\ 0 & 0 & 0 & 0 & 0 & a_2 & 0 & -a_2 & 0 \\ 0 & 0 & 0 & 0 & 0 & 0 & 0 & 0 & a_2 \\ 0 & 0 & 0 & 0 & 0 & 0 & 0 & 0 & 0 \\ 0 & 0 & 0 & 0 & 0 & 0 & 0 & 0 & -a_2 \\ 0 & 0 & 0 & 0 & 0 & 0 & 0 & 0 & 0 \end{pmatrix}. \tag{5.12}$$

- On the other hand we have

$$r := \frac{1}{2}\begin{pmatrix} 0 & 0 & 0 & 0 & 0 & 0 & 0 & 0 & 0 \\ 0 & 0 & 0 & 1 & 0 & 0 & 0 & 0 & 0 \\ 0 & 0 & 0 & 0 & 0 & 0 & 1 & 0 & 0 \\ 0 & -1 & 0 & 0 & 0 & 0 & 0 & 0 & 0 \\ 0 & 0 & 0 & 0 & 0 & 0 & 0 & 0 & 0 \\ 0 & 0 & 0 & 0 & 0 & 0 & 0 & 1 & 0 \\ 0 & 0 & -1 & 0 & 0 & 0 & 0 & 0 & 0 \\ 0 & 0 & 0 & 0 & 0 & -1 & 0 & 0 & 0 \\ 0 & 0 & 0 & 0 & 0 & 0 & 0 & 0 & 0 \end{pmatrix},$$

and

$$r_0 := \frac{1}{2}\begin{pmatrix} 1 & 0 & 0 & 0 & 0 & 0 & 0 & 0 & 0 \\ 0 & 0 & 0 & 0 & 0 & 0 & 0 & 0 & 0 \\ 0 & 0 & 0 & 0 & 0 & 0 & 0 & 0 & 0 \\ 0 & 0 & 0 & 0 & 0 & 0 & 0 & 0 & 0 \\ 0 & 0 & 0 & 0 & 1 & 0 & 0 & 0 & 0 \\ 0 & 0 & 0 & 0 & 0 & 0 & 0 & 0 & 0 \\ 0 & 0 & 0 & 0 & 0 & 0 & 0 & 0 & 0 \\ 0 & 0 & 0 & 0 & 0 & 0 & 0 & 0 & 0 \\ 0 & 0 & 0 & 0 & 0 & 0 & 0 & 0 & 1 \end{pmatrix}.$$

- Now it is easy to verify that $[\,\mathsf{X}\otimes 1, r_0 - r\,] - [\,1\otimes \mathsf{X}, r + r_0\,]$ is equal to (5.12).

We proved that the claim is true for $N = 3$. ■

5.2 The Garnier system

► Garnier systems (1919) are a class of completely integrable systems closely related to mechanical systems with cubic nonlinearities.

- The Newton equations of motion of the Garnier system are defined on $\mathbb{R}^{2N}$, parametrized with coordinates $q := (q_1, \ldots, q_N)^\top$ and $a := (a_1, \ldots, a_N)^\top$. They read

$$\begin{cases} \ddot{q}_i = -\omega_i\, q_i - 2\, q_i \,\langle\, q, a\,\rangle\,, \\ \ddot{a}_i = -\omega_i\, a_i - 2\, a_i \,\langle\, q, a\,\rangle\,, \end{cases} \tag{5.13}$$

  with $i = 1, \ldots, N$. Here $\langle\,\cdot,\cdot\,\rangle$ denotes the standard Euclidean scalar product on $\mathbb{R}^N$ and the ω_i, $i = 1, \ldots, N$, are all distinct real parameters.

- Equations (5.13) define a Hamiltonian system in $\mathbb{R}^{4N}$ equipped with coordinates $(q, a, p, b) := (q_1, \ldots, q_N, a_1, \ldots, a_N, p_1, \ldots, p_N, b_1, \ldots, b_N)$ and Poisson brackets

$$\{\, q_i, b_j \,\} = \delta_{ij}, \qquad \{\, a_i, p_j \,\} = \delta_{ij}. \tag{5.14}$$

  The Hamiltonian is

$$H := \langle\, p, b\,\rangle + \langle\, \Omega\, q, a\,\rangle + \langle\, q, a\,\rangle^2\,, \tag{5.15}$$

where $\Omega := \mathsf{diag}(\omega_1, \dots, \omega_N)$ and the corresponding Hamilton equations are

$$\begin{cases} \dot{q} = \mathsf{grad}_b\, H = p, \\ \dot{a} = \mathsf{grad}_p\, H = b, \\ \dot{p} = -\,\mathsf{grad}_a\, H = -\Omega\, q - 2\, q\, \langle\, q, a\,\rangle\,, \\ \dot{b} = -\,\mathsf{grad}_q\, H = -\Omega\, a - 2\, a\, \langle\, q, a\,\rangle\,. \end{cases} \tag{5.16}$$

► The complete integrability of the vector field (5.16) is established once we have $4N - 2N = 2N$ integrals of motion which are Poisson involutive and independent.

5.2.1 Complete integrability

► In the next claim we give a Lax formulation for equations of motion (5.16). This is defined on the loop algebra $\mathfrak{gl}(N+1, \mathbb{R})[\lambda, \lambda^{-1}]$.

Theorem 5.4

1. *Equations (5.16) admits the following Lax representation on the loop algebra $\mathfrak{gl}(N+1, \mathbb{R})[\lambda, \lambda^{-1}]$:*

$$\dot{\mathsf{X}}(\lambda) = -[\,\mathsf{X}(\lambda), \mathsf{B}(\lambda)\,], \tag{5.17}$$

where the Lax operator is given by

$$\mathsf{X}(\lambda) := \begin{pmatrix} q \otimes a + \Omega & p + \lambda\, q \\ b^\top - \lambda\, a^\top & -\,\langle\, q, a\,\rangle - \lambda^2 \end{pmatrix}, \tag{5.18}$$

where $q \otimes a := q\, a^\top$, while the auxiliary matrix is

$$\mathsf{B}(\lambda) := \begin{pmatrix} 0 & q \\ -a^\top & -\lambda \end{pmatrix}. \tag{5.19}$$

2. *The characteristic polynomial of the Lax operator, $\mathsf{det}(\mathsf{X}(\lambda) - \mu\, 1)$, admits the following representation:*

$$\mathsf{det}(\mathsf{X}(\lambda) - \mu\, 1) = \prod_{j=1}^{N} (\omega_j - \mu) \left(\mu + \lambda^2 + \sum_{i=1}^{N} \frac{F_i + \lambda\, G_i}{\omega_i - \mu} \right),$$

where

$$\begin{aligned} F_i &:= p_i\, b_i + q_i\, a_i (\omega_i + \langle\, q, a\,\rangle) + \sum_{j \neq i} \frac{(q_i\, p_j - q_j\, p_i)(a_i\, b_j - a_j\, b_i)}{\omega_j - \omega_i}, \\ G_i &:= q_i\, b_i - a_i\, p_i, \end{aligned}$$

with $i = 1, \dots, N$. *The functions* $(F_1, \dots, F_N, G_1, \dots, G_N)$ *provide a set of* $2N$ *independent integrals of motion of (5.16). In particular, the Hamiltonian (5.15) is given by*

$$H = \sum_{i=1}^{N} F_i.$$

Proof. We prove only the equivalence between equations (5.16) and the Lax representation (5.17). We have:

$$\mathsf{X}(\lambda)\,\mathsf{B}(\lambda) = \begin{pmatrix} -p \otimes a - \lambda\, q \otimes a & (q \otimes a)\, q + \Omega\, q - \lambda\, p - \lambda^2 q \\ \langle q, a \rangle\, a^\top + \lambda^2 a^\top & \langle b, q \rangle + \lambda^3 \end{pmatrix},$$

and

$$\mathsf{B}(\lambda)\,\mathsf{X}(\lambda) = \begin{pmatrix} q \otimes b - \lambda\, q \otimes a & -\langle q, a \rangle\, q - \lambda^2 q \\ -a^\top (q \otimes a) - a^\top \Omega - \lambda\, b^\top + \lambda^2 a^\top & -\langle a, p \rangle + \lambda^3 \end{pmatrix}.$$

Now recall the vectorial identity

$$\langle q, a \rangle\, a^\top = a^\top (q \otimes a).$$

Therefore,

$$-[\mathsf{X}(\lambda), \mathsf{B}(\lambda)] = \begin{pmatrix} q \otimes b - p \otimes a & -\Omega\, q - 2q\, \langle q, a \rangle + \lambda\, p \\ -a^\top \Omega - 2a^\top \langle q, a \rangle - \lambda\, b^\top & -\langle a, p \rangle - \langle b, q \rangle \end{pmatrix}.$$

Comparing the above matrix with $\dot{\mathsf{X}}(\lambda)$ we get equations of motion (5.16). ■

► The fact that the spectral invariants of the Lax operator (5.17) are Poisson involutive on $\mathbb{R}^{4N}$ equipped with the brackets (5.14) is a consequence of the next Theorem combined with Theorem 4.12.

Theorem 5.5

The Lax operator (5.17) satisfies a linear r*-matrix bracket*

$$\{\, \mathsf{X}(\lambda) \overset{\otimes}{,} \mathsf{X}(\mu) \,\} = [\, \mathsf{X}(\lambda) \otimes 1 + 1 \otimes \mathsf{X}(\mu), r(\lambda - \mu) \,],$$

where

$$r(\lambda - \mu) := \frac{1}{\lambda - \mu} \sum_{i,j=1}^{N+1} \mathsf{E}_{ij} \otimes \mathsf{E}_{ji}.$$

No Proof.

5.3 The rational $\mathfrak{su}(2)$ Gaudin system

► The systems introduced in 1976 by Gaudin and carrying nowadays his name attracted considerable interest among theoretical and mathematical physicists, playing a distinguished role in the realm of integrable systems.

- The Gaudin systems describe completely integrable classical and quantum long-range interacting spin chains.
- Originally the Gaudin system was formulated as a spin system related to the Lie algebra $\mathfrak{sl}(2)$. Later it was realized that one can associate such a system with any semi-simple (complex) Lie algebra $\mathfrak{g}$ and a solution of the corresponding Yang-Baxter equation.
- Depending on the anisotropy of interaction between spins, one distinguishes between XXX, XXZ and XYZ systems. Corresponding Lax matrices are defined on loop algebras and they turn out to depend on the spectral parameter through rational, trigonometric and elliptic functions, respectively.
- Both the classical and the quantum Gaudin systems can be formulated within the r-matrix approach. They admit a linear r-matrix structure. It is worthwhile to mention that Gaudin systems also admit a multi-Hamiltonian formulation.

► In what follows we present the rational (i.e., XXX) Gaudin system in the case of $\mathfrak{g} = \mathfrak{su}(2)$. We will see that its Lax representation can be formulated in terms of a loop algebra given by the direct sum of N identical copies of $\mathfrak{su}(2)[\lambda, \lambda^{-1}]$.

5.3.1 Algebraic setting

► The basic Lie algebra is $\mathfrak{g} = \mathfrak{su}(2)$ ($\simeq \mathfrak{so}(3)$), which is the vector space spanned by traceless and antihermitian matrices.

- We choose the following basis for $\mathfrak{su}(2)$:

$$\mathsf{E}_1 := \frac{1}{2}\begin{pmatrix} 0 & -\mathrm{i} \\ -\mathrm{i} & 0 \end{pmatrix}, \qquad \mathsf{E}_2 := \frac{1}{2}\begin{pmatrix} 0 & -1 \\ 1 & 0 \end{pmatrix}, \qquad \mathsf{E}_3 := \frac{1}{2}\begin{pmatrix} -\mathrm{i} & 0 \\ 0 & \mathrm{i} \end{pmatrix}.$$

  Note that $\mathsf{E}_\alpha^2 = -1/4$ and

$$[\,\mathsf{E}_\alpha, \mathsf{E}_\beta\,] = \varepsilon_{\alpha\beta\gamma}\, \mathsf{E}_\gamma.$$

- The correspondence

$$\mathsf{X} := \left(x^1, x^2, x^3\right)^\top \leftrightarrow \mathsf{X} := \frac{1}{2}\begin{pmatrix} -\mathrm{i}\,x^3 & -\mathrm{i}\,x^1 - x^2 \\ -\mathrm{i}\,x^1 + x^2 & \mathrm{i}\,x^3 \end{pmatrix} = \sum_{\alpha=1}^{3} x^\alpha\, \mathsf{E}_\alpha,$$

is a Lie algebra isomorphism between $(\mathbb{R}^3, \times)$ and $(\mathfrak{su}(2), [\,\cdot,\cdot\,])$ (here $\times$ stands for the vector product and $[\,\cdot,\cdot\,]$ for the standard commutator of matrices). The above isomorphism allows us to identify vectors from $\mathbb{R}^3$ with matrices from $\mathfrak{su}(2)$.

- We equip $\mathfrak{su}(2)$ with the scalar product $\langle\,\cdot,\cdot\,\rangle$ induced from $\mathbb{R}^3$, namely

$$\langle\, \mathsf{X}, \mathsf{Y}\,\rangle := -2\,\mathsf{Trace}\,(\mathsf{X}\,\mathsf{Y}), \qquad \mathsf{X}, \mathsf{Y} \in \mathfrak{su}(2).$$

  This scalar product allows us to identify the dual space $\mathfrak{su}(2)^*$ with $\mathfrak{su}(2)$, so that the coadjoint action of the algebra becomes the usual Lie bracket with minus, i.e. $\mathrm{ad}^*_{\mathsf{Y}}\,\mathsf{X} = [\,\mathsf{X}, \mathsf{Y}\,] = -\mathrm{ad}_{\mathsf{Y}}\,\mathsf{X}$, with $\mathsf{X}, \mathsf{Y} \in \mathfrak{su}(2)$ (note that this identification allows us to not distinguish between $\langle\,\cdot,\cdot\,\rangle$ and $\langle\,\cdot\,|\,\cdot\,\rangle$).

- The matrix multiplication and the commutator in $\mathfrak{su}(2)$ are related by the following formula:

$$\mathsf{X}\,\mathsf{Y} = -\frac{1}{4}\langle\,\mathsf{X}, \mathsf{Y}\,\rangle \mathbb{1} + \frac{1}{2}[\,\mathsf{X}, \mathsf{Y}\,], \qquad \mathsf{X}, \mathsf{Y} \in \mathfrak{su}(2). \tag{5.20}$$

  In particular, if $\langle\, \mathsf{X}, \mathsf{Y}\,\rangle = 0$ then $\mathsf{X}\,\mathsf{Y} + \mathsf{Y}\,\mathsf{X} = 0$.

► The global Lie algebra of the system is

$$\mathfrak{G} := \bigoplus^{N} \mathfrak{su}(2) \simeq \mathbb{R}^{3N}.$$

Each copy of the basic Lie algebra $\mathfrak{su}(2)$ is associated with a single member of the system.

5.3.2 *Definition of the phase space and equations of motion*

► We now define the phase space of the system and its equations of motion.

- We denote by $\mathsf{X}_i := \left(x_i^1, x_i^2, x_i^3\right)^\top, i = 1, \ldots, N$, the coordinate functions (in the basis E_α) on the i-th copy of $\mathfrak{su}(2)^*$. The global phase space is $\mathfrak{G}^*$, which has dimension $3N$, and any X_i defined on the i-th summand of $\mathfrak{G}^* \simeq \mathfrak{G}$ can be expressed as

$$\mathsf{X}_i = \sum_{\alpha=1}^{3} x_i^\alpha\, \mathsf{E}_\alpha, \qquad i = 1, \ldots, N. \tag{5.21}$$

- The phase space becomes a Lie-Poisson manifold if we equip $\mathfrak{G}^*$ with the Lie-Poisson bracket

$$\left\{x_i^\alpha, x_j^\beta\right\} = -\delta_{ij}\,\varepsilon_{\alpha\beta\gamma}\,x_i^\gamma, \qquad i, j = 1, \ldots, N. \tag{5.22}$$

- The bracket (5.22) is degenerate since the rank of the Poisson structure is $2N$. Indeed, it possesses $3N - 2N = N$ independent Casimirs defined by

$$C_i := \frac{1}{2}\langle \mathsf{X}_i, \mathsf{X}_i \rangle, \qquad i = 1, \ldots, N. \tag{5.23}$$

  Fixing their values we get a foliation of $\mathfrak{G}^*$,which is given by the union of N two-dimensional spheres.

- We define the Hamiltonian of the rational $\mathfrak{su}(2)$ Gaudin system as

$$H := \frac{1}{2}\sum_{\substack{i,j=1\\ i\neq j}}^{N} \langle \mathsf{X}_i, \mathsf{X}_j \rangle + \sum_{i=1}^{N} \omega_i \langle \mathsf{P}, \mathsf{X}_i \rangle. \tag{5.24}$$

  The first term describes an interaction between vectors X_i and X_j (which are interpreted as $\mathfrak{su}(2)$ spins in the quantum case) while the second term is an interaction of vectors X_i with a constant external field $\mathsf{P} \in \mathbb{R}^3$. The distinct parameters ω_i (which can be either complex or real) are parameters of the system.

- The Hamiltonian vector field corresponding to H w.r.t. the bracket (5.22) is given by

$$\dot{\mathsf{X}}_i = [\, \mathsf{grad}_{\mathsf{X}_i} H, \mathsf{X}_i \,] = \left[\omega_i \mathsf{P} + \sum_{j=1}^{N} \mathsf{X}_j \,,\, \mathsf{X}_i \right], \qquad i = 1, \ldots, N. \tag{5.25}$$

► The complete integrability of the vector field (5.25) is established once we have $3N - N = 2N$ integrals of motion, $(F_1, \ldots, F_{2N})$, which are Poisson involutive and independent. Among these $2N$ functions we already know N Casimirs and the Hamiltonian (5.24).

5.3.3 *Complete integrability*

► To prove the complete integrability of the vector field (5.25) we first show that (5.25) is equivalent to a Lax equation written on the loop algebra $\mathfrak{G}[\lambda, \lambda^{-1}]$, $\lambda \in \mathbb{C}$. This will guarantee the existence of $2N$ integrals of motion. Then we will prove that the Lax operator admits a linear r-matrix structure, thus showing the Poisson involutivity of the integrals of motion.

Theorem 5.6

1. *Equations (5.25) admits the following Lax representation on the loop algebra* $\mathfrak{G}[\lambda, \lambda^{-1}]$*:*

$$\dot{\mathsf{X}}(\lambda) = [\mathsf{X}(\lambda), \mathsf{B}_-(\lambda)] = -[\mathsf{X}(\lambda), \mathsf{B}_+(\lambda)], \tag{5.26}$$

where the Lax operator is given by

$$\mathsf{X}(\lambda) := \mathsf{P} + \sum_{i=1}^{N} \frac{\mathsf{X}_i}{\lambda - \omega_i}, \tag{5.27}$$

while the auxiliary matrices are

$$\mathsf{B}_-(\lambda) := \sum_{i=1}^{N} \frac{\omega_i \mathsf{X}_i}{\lambda - \omega_i}, \qquad \mathsf{B}_+(\lambda) := \lambda \mathsf{P} + \sum_{i=1}^{N} \mathsf{X}_i.$$

2. *The characteristic polynomial of the Lax operator,* $\det(\mathsf{X}(\lambda) - \mu \mathbb{1})$, *admits the following representation:*

$$\det(\mathsf{X}(\lambda) - \mu \mathbb{1}) = \mu^2 + \frac{1}{4}\langle \mathsf{P}, \mathsf{P} \rangle + \frac{1}{2}\sum_{i=1}^{N} \left(\frac{H_i}{\lambda - \omega_i} + \frac{C_i}{(\lambda - \omega_i)^2} \right), \tag{5.28}$$

where the functions C_i are the Casimirs (5.23), whereas the functions H_i are

$$H_i := \langle \mathsf{P}, \mathsf{X}_i \rangle + \sum_{\substack{j=1 \\ j \neq i}}^{N} \frac{\langle \mathsf{X}_i, \mathsf{X}_j \rangle}{\omega_i - \omega_j}, \qquad i = 1, \dots, N.$$

The functions $(C_1, \dots, C_N, H_1, \dots, H_N)$ provide a set of $2N$ independent integrals of motion of (5.25). In particular, the Hamiltonian (5.24) is given by

$$H = \sum_{i=1}^{N} \omega_i H_i. \tag{5.29}$$

Proof. We sketch a proof for both claims.

1. Let us prove the claim for the auxiliary matrix $\mathsf{B}_-(\lambda)$:

$$\begin{aligned} \dot{\mathsf{X}}(\lambda) &= \sum_{i=1}^{N} \frac{\dot{\mathsf{X}}_i}{\lambda - \omega_i} = \left[\mathsf{P} + \sum_{i=1}^{N} \frac{\mathsf{X}_i}{\lambda - \omega_i}, \sum_{j=1}^{N} \frac{\omega_j \mathsf{X}_j}{\lambda - \omega_j} \right] \\ &= \sum_{i=1}^{N} \frac{\omega_i}{\lambda - \omega_i} [\mathsf{P}, \mathsf{X}_i] + \frac{1}{2} \sum_{i,j=1}^{N} \frac{\omega_j - \omega_i}{(\lambda - \omega_i)(\lambda - \omega_j)} [\mathsf{X}_i, \mathsf{X}_j]. \end{aligned}$$

   Equations of motion (5.25) for X_k are now recovered taking the residues of the above Lax equation at poles $\lambda = \omega_k$, $k = 1, \dots, N$.

2. Since $\mathsf{X} \in \mathfrak{G}[\lambda, \lambda^{-1}]$ we immediately find the following expression:

$$\det(\mathsf{X}(\lambda) - \mu \mathbb{1}) = \mu^2 + \frac{1}{4} \langle \mathsf{X}(\lambda), \mathsf{X}(\lambda) \rangle .$$

Now, taking into account (5.27) a straightforward computation leads to (5.28). Relation (5.29) is obvious.

The Theorem is proved. ∎

► The fact that the functions $(C_1, \dots, C_N, H_1, \dots, H_N)$ are Poisson involutive on $\mathfrak{G}^*$ is a consequence of the next Theorem combined with Theorem 4.12.

Theorem 5.7

The Lax operator (5.27) satisfies a linear r-matrix bracket

$$\{\mathsf{X}(\lambda) \overset{\otimes}{,} \mathsf{X}(\mu)\} = [\mathsf{X}(\lambda) \otimes 1 + 1 \otimes \mathsf{X}(\mu), r(\lambda - \mu)], \tag{5.30}$$

where the r-matrix given by

$$r(\lambda - \mu) := -\frac{1}{\lambda - \mu}(\mathsf{E}_1 \otimes \mathsf{E}_1 + \mathsf{E}_2 \otimes \mathsf{E}_2 + \mathsf{E}_3 \otimes \mathsf{E}_3). \tag{5.31}$$

Proof. Thanks to the locality of the Lie-Poisson bracket (5.22) it is sufficient to verify (5.30) only for a given site of the chain, say the *i*-th. We also observe that the non-dynamical part P of the Lax matrix does not play any role in such a computation. We proceed by steps (we use summation over repeated greek indices).

- To verify (5.30) we compute

$$\begin{aligned}\left\{\frac{x_i^\alpha}{\lambda - \omega_i}\mathsf{E}_\alpha \otimes 1, \frac{x_i^\beta}{\mu - \omega_i} 1 \otimes \mathsf{E}_\beta\right\} &= -\frac{\varepsilon_{\alpha\beta\gamma} x_i^\gamma}{(\lambda - \omega_i)(\mu - \omega_i)}\mathsf{E}_\alpha \otimes \mathsf{E}_\beta \\ &= -\frac{x_i^1\, \mathsf{E}_{23} + x_i^2\, \mathsf{E}_{31} + x_i^3\, \mathsf{E}_{12}}{(\lambda - \omega_i)(\mu - \omega_i)}.\end{aligned}$$

  where $\mathsf{E}_{\alpha\beta} := \mathsf{E}_\alpha \otimes \mathsf{E}_\beta - \mathsf{E}_\beta \otimes \mathsf{E}_\alpha$.

- On the other hand we have:

$$\begin{aligned}&-\frac{1}{\lambda - \mu}\left[\frac{x_i^\alpha}{\lambda - \omega_i}\mathsf{E}_\alpha \otimes 1, \mathsf{E}_1 \otimes \mathsf{E}_1 + \mathsf{E}_2 \otimes \mathsf{E}_2 + \mathsf{E}_3 \otimes \mathsf{E}_3\right] \\ &\quad = -\frac{x_i^1}{(\lambda - \mu)(\lambda - \omega_i)}[\mathsf{E}_1 \otimes 1, \mathsf{E}_1 \otimes \mathsf{E}_1 + \mathsf{E}_2 \otimes \mathsf{E}_2 + \mathsf{E}_3 \otimes \mathsf{E}_3] \\ &\quad\quad -\frac{x_i^2}{(\lambda - \mu)(\lambda - \omega_i)}[\mathsf{E}_2 \otimes 1, \mathsf{E}_1 \otimes \mathsf{E}_1 + \mathsf{E}_2 \otimes \mathsf{E}_2 + \mathsf{E}_3 \otimes \mathsf{E}_3] \\ &\quad\quad -\frac{x_i^3}{(\lambda - \mu)(\lambda - \omega_i)}[\mathsf{E}_3 \otimes 1, \mathsf{E}_1 \otimes \mathsf{E}_1 + \mathsf{E}_2 \otimes \mathsf{E}_2 + \mathsf{E}_3 \otimes \mathsf{E}_3] \\ &\quad = \frac{x_i^1\, \mathsf{E}_{23} + x_i^2\, \mathsf{E}_{31} + x_i^3\, \mathsf{E}_{12}}{(\lambda - \mu)(\lambda - \omega_i)},\end{aligned}$$

where we used $[\mathsf{E}_\alpha \otimes 1, \mathsf{E}_\beta \otimes \mathsf{E}_\beta] = [\mathsf{E}_\alpha, \mathsf{E}_\beta] \otimes \mathsf{E}_\beta = \varepsilon_{\alpha\beta\gamma}\mathsf{E}_\gamma \otimes \mathsf{E}_\beta$.

- Similarly,

$$-\frac{1}{\lambda-\mu}\left[\frac{x_i^\alpha}{\mu-\omega_i} 1 \otimes \mathsf{E}_\alpha, \mathsf{E}_1 \otimes \mathsf{E}_1 + \mathsf{E}_2 \otimes \mathsf{E}_2 + \mathsf{E}_3 \otimes \mathsf{E}_3\right]$$
$$= -\frac{x_i^1\,\mathsf{E}_{23} + x_i^2\,\mathsf{E}_{31} + x_i^3\,\mathsf{E}_{12}}{(\lambda-\mu)(\mu-\omega_i)}.$$

- Now, taking into account the last formulas, we have

$$-\frac{1}{\lambda-\mu}\left[\frac{x_i^\alpha}{\lambda-\omega_i}\mathsf{E}_\alpha \otimes 1 + \frac{x_i^\alpha}{\mu-\omega_i} 1 \otimes \mathsf{E}_\alpha, \mathsf{E}_1 \otimes \mathsf{E}_1 + \mathsf{E}_2 \otimes \mathsf{E}_2 + \mathsf{E}_3 \otimes \mathsf{E}_3\right]$$
$$= \frac{1}{\lambda-\mu}\left(\frac{1}{\lambda-\omega_i} - \frac{1}{\mu-\omega_i}\right)\left(x_i^1\,\mathsf{E}_{23} + x_i^2\,\mathsf{E}_{31} + x_i^3\,\mathsf{E}_{12}\right)$$
$$= -\frac{x_i^1\,\mathsf{E}_{23} + x_i^2\,\mathsf{E}_{31} + x_i^3\,\mathsf{E}_{12}}{(\lambda-\omega_i)(\mu-\omega_i)}.$$

The Theorem is proved. ■

► One can directly verify that the r-matrix (5.31) satisfies the Yang-Baxter equation (4.42).

Example 5.2 (*A two-body Gaudin system*)

Consider the vector field (5.25) with $N = 2$. To simplify the notation we set $\mathsf{X}_1 := (x_1, x_2, x_3)$ and $\mathsf{X}_2 := (y_1, y_2, y_3)$. Moreover we fix $P := (0, 0, 1)$. We thus obtain the following system of differential equations:

$$\begin{cases} \dot{x}_1 = x_2\,y_3 - x_3\,y_2 + \omega_1\,x_2, \\ \dot{x}_2 = x_3\,y_1 - x_1\,y_3 - \omega_1\,x_1, \\ \dot{x}_3 = x_1\,y_2 - x_2\,y_1, \\ \dot{y}_1 = y_2\,x_3 - y_3\,x_2 + \omega_2\,y_2, \\ \dot{y}_2 = y_3\,x_1 - y_1\,x_3 - \omega_2\,y_1, \\ \dot{y}_3 = y_1\,x_2 - y_2\,x_1. \end{cases} \tag{5.32}$$

The system (5.32) has the following four independent integrals of motion:

$$C_1 = x_2^2 + x_2^2 + x_3^2, \qquad C_2 = y_1^2 + y_2^2 + y_3^2,$$
$$H_1 = x_3 + \frac{x_1\,y_1 + x_2\,y_2 + x_3\,y_3}{\omega_1 - \omega_2}, \qquad H_2 = y_3 + \frac{x_1\,y_1 + x_2\,y_2 + x_3\,y_3}{\omega_2 - \omega_1},$$

which are Poisson involutive w.r.t. the Lie-Poisson bracket of $(\mathfrak{su}(2) \oplus \mathfrak{su}(2))^*$.

5.3.4 *Algebraic contractions of the phase space*

► We now show that a special algebraic procedure, called *Lie algebra contraction*, performed on the phase space $\mathfrak{G}^*$ allows us to obtain interesting integrable systems from the Gaudin system. A simple instance of these systems is given by the famous *Lagrange top*.

► In the next claim we see an instance of a quite general algebraic procedure called *Lie algebra contraction*. Probably the most famous Lie algebra contraction is illustrated by the mechanism which relates relativistic and classical mechanics with their underlying Lorentz and Galilei algebras. Another well-known example is the limiting process from quantum mechanics to classical mechanics.

Theorem 5.8

Consider the Lie-Poisson algebra $\mathfrak{G}^* \simeq \mathbb{R}^{3N}$ *equipped with the bracket (5.22) and parametrized by* $\mathsf{X}_1, \ldots, \mathsf{X}_N$. *On* $\mathfrak{G}^*$ *consider the isomorphism defined by*

$$\mathsf{Y}_i := \varepsilon^i \sum_{j=1}^{N} \nu_j^i \mathsf{X}_j, \qquad i = 0, \ldots, N-1, \tag{5.33}$$

with distinct parameters $\nu_j \in \mathbb{C}$ *and* $0 < \varepsilon \leqslant 1$. *Here* $\mathsf{Y}_i := \left(y_i^1, y_i^2, y_i^3\right)^\top$.

1. *Under the map (5.33) the Lie-Poisson bracket (5.22) induces for* $\varepsilon \to 0$ *a Lie-Poisson bracket for the coordinates* $\mathsf{Y}_0, \ldots, \mathsf{Y}_{N-1}$ *given by*

$$\left\{ y_i^\alpha, y_j^\beta \right\}_N = \begin{cases} -\varepsilon_{\alpha\beta\gamma}\, y_{i+j}^\gamma & i+j < N, \\ 0 & i+j \geqslant N, \end{cases} \tag{5.34}$$

   with $i, j = 0, \ldots, N-1$. *We denote by* $\mathfrak{G}_N^* \simeq \mathbb{R}^{3N}$ *the Lie-Poisson algebra defined by the bracket (5.34).*

2. *The bracket (5.34) has rank* $2N$. *It admits* N *Casimirs given by*

$$G_k := \frac{1}{2} \sum_{i=k}^{N-1} \langle \mathsf{Y}_i, \mathsf{Y}_{N+k-i-1} \rangle, \qquad k = 0, \ldots, N-1. \tag{5.35}$$

Proof. We prove both claims.

1. Using (5.22) and (5.33) we get:

$$\begin{aligned} \left\{ y_i^\alpha, y_j^\beta \right\}_\varepsilon &= \varepsilon^{i+j} \sum_{n,m=1}^{N} \nu_n^i \, \nu_m^j \left\{ x_n^\alpha, x_m^\beta \right\} \\ &= -\varepsilon_{\alpha\beta\gamma}\, \varepsilon^{i+j} \sum_{n=1}^{N} \nu_n^{i+j} \, x_n^\gamma \\ &= \begin{cases} -\varepsilon_{\alpha\beta\gamma}\, y_{i+j}^\gamma & i+j < N, \\ O(\varepsilon) & i+j \geqslant N. \end{cases} \end{aligned}$$

The limit $\varepsilon \to 0$ leads to (5.34). It is easy to check that (5.34) defines indeed a Lie algebra structure.

2. A direct computation shows that $\{G_k, y_i^\alpha\}_N = 0$ for any y_i^α and for all $k = 0, \ldots, N-1$. Moreover the N functions G_k are independent.

The Theorem is proved. ■

▶ Remarks:

- Note that $\mathfrak{G}_N^*$ does not have a direct sum structure as $\mathfrak{G}^*$, which is simply the direct sum of N identical copies of $\mathfrak{su}(2)^*$.
- Fix $N = 2$. The map (5.33) is

$$\mathsf{Y}_0 := \mathsf{X}_1 + \mathsf{X}_2, \qquad \mathsf{Y}_1 := \varepsilon\,(\nu_1 \mathsf{X}_1 + \nu_2 \mathsf{X}_2).$$

The Lie-Poisson bracket on $\mathfrak{G}_2^*$ reads

$$\left\{y_0^\alpha, y_0^\beta\right\}_2 = -\varepsilon_{\alpha\beta\gamma}\, y_0^\gamma, \qquad \left\{y_0^\alpha, y_1^\beta\right\}_2 = -\varepsilon_{\alpha\beta\gamma}\, y_1^\gamma, \qquad \left\{y_1^\alpha, y_1^\beta\right\}_2 = 0, \tag{5.36}$$

which is the Lie-Poisson bracket of $\mathfrak{e}(3)^* \simeq \mathfrak{su}(2)^* \oplus_s \mathbb{R}^3$. Its Casimirs are

$$G_0 := \langle \mathsf{Y}_0, \mathsf{Y}_1 \rangle, \qquad G_1 := \frac{1}{2}\langle \mathsf{Y}_1, \mathsf{Y}_1 \rangle.$$

5.3.5 From rational $\mathfrak{su}(2)$ Gaudin systems to Lagrange top systems

▶ Our task is now to apply the map (5.33), in the contraction limit $\varepsilon \to 0$, to the Lax operator (5.27), in order to get a new Lax operator governing integrable systems defined on $\mathfrak{G}_N^*$.

▶ We give the following claim.

Theorem 5.9

Consider a rational $\mathfrak{su}(2)$ Gaudin system defined on $\mathfrak{G}^*[\lambda, \lambda^{-1}]$.

1. *Under the map (5.33) the Lax operator (5.27), with $\omega_i = \varepsilon\,\nu_i$, $i = 1, \ldots, N$, defines for $\varepsilon \to 0$ a Lax operator on $\mathfrak{G}_N^*[\lambda, \lambda^{-1}]$ given by*

$$\mathsf{Y}(\lambda) := \mathsf{P} + \sum_{i=0}^{N-1} \frac{\mathsf{Y}_i}{\lambda^{i+1}}. \tag{5.37}$$

Correspondingly, the Lax equation (5.26) turns into

$$\dot{\mathsf{Y}}(\lambda) = [\,\mathsf{Y}(\lambda), \mathsf{D}_-(\lambda)\,] = -\,[\,\mathsf{Y}(\lambda), \mathsf{D}_+(\lambda)\,], \tag{5.38}$$

with

$$\mathsf{D}_-(\lambda) := \sum_{i=1}^{N-1} \frac{\mathsf{Y}_i}{\lambda^i}, \qquad \mathsf{D}_+(\lambda) := \lambda\,\mathsf{P} + \mathsf{Y}_0. \tag{5.39}$$

The Lax equation (5.38) is equivalent to the following equations of motion:

$$\dot{\mathsf{Y}}_i = [\,\mathsf{Y}_0, \mathsf{Y}_i\,] + [\,\mathsf{P}, \mathsf{Y}_{i+1}\,], \qquad i = 0, \ldots, N-1, \tag{5.40}$$

with $\mathsf{Y}_N \equiv 0$.

2. *The characteristic polynomial of the Lax operator,* $\det(\mathsf{Y}(\lambda) - \mu\,\mathbb{1})$*, admits the following representation:*

$$\det(\mathsf{Y}(\lambda) - \mu\,\mathbb{1}) = \mu^2 + \frac{1}{4}\langle\,\mathsf{P}, \mathsf{P}\,\rangle + \frac{1}{2}\sum_{k=0}^{N-1}\left(\frac{F_k}{\lambda^{k+1}} + \frac{G_k}{\lambda^{k+N+1}}\right),$$

where the functions G_k *are the Casimirs (5.35), whereas the functions* F_k *are*

$$F_k := \langle\,\mathsf{P}, \mathsf{Y}_k\,\rangle + \frac{1}{2}\sum_{i=0}^{k-1}\langle\,\mathsf{Y}_i, \mathsf{Y}_{k-i-1}\,\rangle.$$

The functions $(G_0, \ldots, G_{N-1}, F_0, \ldots, F_{N-1})$ *provide a set of* $2N$ *independent integrals of motion of (5.40). In particular, the vector field (5.40) is Hamiltonian w.r.t. (5.34) with Hamiltonian* F_1.

3. *The Lax operator (5.37) satisfies the linear* r*-matrix bracket (5.30) with the same* r*-matrix (5.31). Therefore the functions* $(G_0, \ldots, G_{N-1}, F_0, \ldots, F_{N-1})$ *are Poisson involutive on* $\mathfrak{G}_N^*$.

Proof. We prove only the first part of the first claim. By applying the map (5.33) and the coalescence $\omega_i = \varepsilon\,\nu_i$ on (5.27) we get

$$\mathsf{X}(\lambda) = \mathsf{P} + \sum_{j=1}^{N}\frac{\mathsf{X}_j}{\lambda - \varepsilon\,\nu_j} = \mathsf{P} + \frac{1}{\lambda}\sum_{j=1}^{N}\sum_{i=0}^{N-1}\left(\frac{\varepsilon\,\nu_j}{\lambda}\right)^i \mathsf{X}_j + O(\varepsilon) \xrightarrow{\varepsilon\to 0} \mathsf{Y}(\lambda).$$

Similarly for the auxiliary matrices (5.39):

$$\mathsf{B}_-(\lambda) = \sum_{j=1}^{N}\frac{\varepsilon\,\nu_j\mathsf{X}_j}{\lambda - \varepsilon\,\nu_j} = \sum_{j=1}^{N}\sum_{i=0}^{N-2}\left(\frac{\varepsilon\,\nu_j}{\lambda}\right)^{i+1}\mathsf{X}_j + O(\varepsilon) \xrightarrow{\varepsilon\to 0} \sum_{i=0}^{N-2}\frac{\mathsf{Y}_{i+1}}{\lambda^{i+1}} = \mathsf{D}_-(\lambda),$$

and

$$\mathsf{B}_+(\lambda) = \lambda\,\mathsf{P} + \sum_{i=1}^{N}\mathsf{X}_i = \mathsf{D}_+(\lambda),$$

as desired. ∎

► The simplest case $N = 2$ in Theorem 5.9 corresponds to the famous completely integrable Lagrange top system (described in the rest frame). Such system is a special integrable case of Kirchhoff rigid body equations (see Examples 2.14 and 3.3).

- The Lagrange case of the rigid body motion around a fixed point in a constant gravity field is characterized by the following data: the body is rotationally symmetric w.r.t. the third coordinate axis and the fixed point lies on the symmetry axis.
- The system is Hamiltonian w.r.t. $\mathfrak{e}(3)^* \simeq \mathfrak{G}_2^*$ (see (5.36)) with Hamiltonian

$$F_1 := \langle \mathsf{P}, \mathsf{Y}_1 \rangle + \frac{1}{2} \langle \mathsf{Y}_0, \mathsf{Y}_0 \rangle.$$

  This gives the equations of motion

$$\begin{cases} \dot{\mathsf{Y}}_0 = [\mathsf{P}, \mathsf{Y}_1], \\ \dot{\mathsf{Y}}_1 = [\mathsf{Y}_0, \mathsf{Y}_1], \end{cases}$$

  where $\mathsf{Y}_0 \in \mathbb{R}^3$ is the vector of kinetic momentum of the body, $\mathsf{Y}_1 \in \mathbb{R}^3$ is the vector pointing from the fixed point to the center of mass of the body and P is the constant vector along the gravity field.
- An external observer is interested in the motion of the symmetry axis of the top on the level sets of the Casimir $G_1 := \langle \mathsf{Y}_1, \mathsf{Y}_1 \rangle / 2$. Trajectories turn out to be parametrized in terms of elliptic functions of time.

6

The Problem of Integrable Discretization

6.1 Introduction

► Let v be a smooth vector field defined on a n-dimensional smooth manifold $\mathcal{M}$.

- In a neighborhood $\mathcal{U} \subset \mathcal{M}$, coordinatized by $x := (x_1, \ldots, x_n)$, we have the local expression

$$\mathsf{v} = \sum_{i=1}^{n} f_i(x) \frac{\partial}{\partial x_i}, \qquad f_i \in \mathscr{F}(\mathcal{U}).$$

- We know that v allows us to define, at least locally, a continuous time dynamical system defined in terms of the initial value problem

$$\begin{cases} \dot{x} = f(x), \\ x(0) = x_0. \end{cases} \tag{6.1}$$

- Indeed, given a smooth vector field v on $\mathcal{M}$ we can find for any $p \in \mathcal{M}$ a coordinate chart $\mathcal{U}$ of p, with coordinates $x := (x_1, \ldots, x_n)$, an open subset $\mathcal{U}' \subseteq \mathcal{U}$ and $\varepsilon > 0$, such that the solution $(t,p) \mapsto x(t,p)$ of (6.1) is uniquely defined for $(t,p) \in \mathcal{I}_\varepsilon \times \mathcal{U}'$, where $\mathcal{I}_\varepsilon := \{t \in \mathbb{R} : |t| < \varepsilon\}$. The map

$$\Phi_t : \mathcal{I}_\varepsilon \times \mathcal{U}' \to \mathcal{U} : (t,p) \mapsto \Phi_t(p) := x(t,p)$$

  is the local flow of v. For a fixed $t \in \mathcal{I}_\varepsilon$ the map Φ_t is a local one-parameter Lie groups of diffeomorphism from $\mathcal{U}'$ to $\Phi_t(\mathcal{U}')$.

- Recall that the component $f_i(x)$ is the i-th component of

$$f(x) := \left.\frac{\mathrm{d}}{\mathrm{d}t}\right|_{t=0} \Phi_t(x).$$

► The problem of numerical integration of ODEs is about constructing good numerical approximations of a flow map Φ_t for a given vector field v. Such approximations provide a time discretization of the continuous dynamical system. To assure the coincidence of the qualitative properties of the discretized models with that of the continuous ones is one of the central ideas of numerical analysis. The importance of numerical integration lies in the following observations:

- To extract quantitative information from the models described by ODEs, it is often necessary to solve them numerically with the help of discretization methods.

- It is crucial to assure that the discretized models exhibit the same (or some of the same) qualitative features of the dynamics as their continuous counterparts. For example, existence of integrals of motion, preservation of the Poisson structure.

▶ We now give a formulation of the *problem of integrable discretization* (for finite-dimensional Hamiltonian systems). To do so we consider a Hamiltonian flow Φ_t defined on a Poisson manifold $(\mathcal{M}, \{\cdot,\cdot\})$ which admits sufficiently many independent and Poisson involutive integrals of motion.

- Let ε be a small positive parameter, called *time-step*.
- The problem consists in finding a one-parameter family of diffeomorphisms $x \mapsto \widetilde{x} := \varphi_\varepsilon(x)$, depending smoothly on ε, such that:

  1. The flow Φ_t is approximated in the following sense:

  $$\widetilde{x} := \varphi_\varepsilon(x) = x + \varepsilon f(x) + O(\varepsilon^2). \tag{6.2}$$

     Therefore $\varphi_0(x) = x$.
  2. φ_ε is a one-parameter family of Poisson maps w.r.t. $\{\cdot,\cdot\}$ or w.r.t. some smooth ε-deformation of $\{\cdot,\cdot\}$.
  3. The maps φ_ε admit the necessary number of independent and Poisson involutive integrals of motion which approximate the continuous integrals of motion. Such integrals of motion can be smooth ε-deformations of the continuous ones.

If such maps exist they provide *completely integrable discretizations* of Φ_t.

▶ Remarks:

- In many concrete cases the parameter ε does not need to be small, i.e., integrability properties of the maps hold true for any $\varepsilon \in \mathbb{R}$.
- In general, the construction of completely integrable discretizations is not algorithmic and systematic. There are indeed many possible and different ways to discretize a given system of ODEs.
- Discretizations can appear also in implicit form,

  $$\frac{\widetilde{x} - x}{\varepsilon} = \varphi_\varepsilon(x, \widetilde{x}).$$

  The "Implicit function Theorem" assures the local solvability of such equations, and condition (6.2) is satisfied if $\varphi_0(x, \widetilde{x}) = f(x)$.

- The general motivation of completely integrable discretizations is the actual possibility of applying completely integrable maps $x \mapsto \widetilde{x} := \varphi_\varepsilon(x)$ for actual numerical computations. Indeed, they guarantee a controlled numerical implementation. Generically, this is not the case if the map does not preserve the characteristic features of the continuous system: the long-time behavior may be also chaotic.

6.2 An integrable discretization of the Euler top

► The task of to present Section is to introduce an integrable discretization of a prototypical continuous integrable system, the three-dimensional Euler top. The discretization scheme we consider goes under the name of *Kahan-Hirota-Kimura discretization* and applies to any quadratic vector field. Remarkably such a scheme preserves the complete integrability of the continuous system in many cases of interest, even though its integrability mechanism (if exists) is presently not known.

6.2.1 *The Kahan-Hirota-Kimura discretization*

► We here define the Kahan-Hirota-Kimura scheme.

- Consider a system of quadratic ODEs for $x : \mathbb{R} \to \mathbb{R}^n$:

$$\dot{x} = f(x) := q(x) + h\,x + c, \tag{6.3}$$

where each component of $q : \mathbb{R}^n \to \mathbb{R}^n$ is a quadratic form, $h \in \mathbf{GL}(n, \mathbb{R})$ and $c \in \mathbb{R}^n$.

- System (6.3) is discretized by the following system of difference equations:

$$\frac{\widetilde{x} - x}{\varepsilon} = q(x, \widetilde{x}) + \frac{1}{2}h(x + \widetilde{x}) + c, \tag{6.4}$$

where

$$q(x, \widetilde{x}) := \frac{1}{2}\left(q(x + \widetilde{x}) - q(x) - q(\widetilde{x})\right)$$

is the symmetric bilinear form corresponding to the quadratic form q. Here we use the following notational convention: for a sequence $x : \mathbb{Z} \to \mathbb{R}^n$ we write x for x_k and $\widetilde{x}$ for x_{k+1}. This allows us to get rid of iteration indices $k \mapsto k+1$ in favor of indices labeling the components of x.

► Remarks:

- System (6.4) is linear w.r.t. x and $\widetilde{x}$ and therefore defines an explicit rational map $x \mapsto \widetilde{x} = \varphi_\varepsilon(x)$, which approximates the time-ε-shift along the solutions of the original system (6.3), i.e., $x_k \approx x(k\,\varepsilon)$.

- System (6.4) is invariant under the interchange $x \leftrightarrow \widetilde{x}$ with the simultaneous sign inversion $\varepsilon \mapsto -\varepsilon$. This implies that the discretization is *reversible*: $\varphi_\varepsilon^{-1} = \varphi_{-\varepsilon}$. Therefore, the map φ_ε is birational.

► We now list some remarkable known properties of the Kahan-Hirota-Kimura scheme:

- The scheme (6.4) can be written as

$$\widetilde{x} = x + \varepsilon \left(1 + \frac{\varepsilon}{2} \frac{\partial f}{\partial x} \right)^{-1} f(x).$$

  In particular the approximation (6.2) holds true.

- If $F \in \mathscr{F}(\mathbb{R}^n)$ is an integral of motion of (6.3) then

$$F(\widetilde{x}) = F(x) + O(\varepsilon^3).$$

  Proof. Recall that the condition that F is an integral of motion for (6.3) is equivalent to

$$\sum_{i=1}^{n} \frac{\partial F}{\partial x_i} f_i(x) = 0. \tag{6.5}$$

  If we take the Lie derivative of (6.5) along v we get

$$\begin{aligned} 0 &= \sum_{j=1}^{n} \frac{\partial}{\partial x_j} \left(\frac{\partial F}{\partial x_i} f_i(x) \right) f_j(x) \\ &= \sum_{i,j=1}^{n} \frac{\partial^2 F}{\partial x_i \partial x_j} f_i(x) f_j(x) + \sum_{i,j=1}^{n} \frac{\partial F}{\partial x_i} \frac{\partial f_i}{\partial x_j} f_j(x). \end{aligned} \tag{6.6}$$

  We now compute

$$\begin{aligned} F(\widetilde{x}) &= F\left(x + \varepsilon f(x) + \frac{\varepsilon^2}{2} \frac{\partial f}{\partial x} f(x) + O(\varepsilon^3) \right) \\ &= F(x) + \varepsilon \sum_{i=1}^{n} \frac{\partial F}{\partial x_i} f_i(x) \\ &\quad + \frac{\varepsilon^2}{2} \left(\sum_{i=1}^{n} \frac{\partial F}{\partial x_i} \left(\frac{\partial f}{\partial x} f(x) \right)_i + \sum_{i,j=1}^{n} \frac{\partial^2 F}{\partial x_i \partial x_j} f_i(x) f_j(x) \right) + O(\varepsilon^3), \end{aligned}$$

  where

$$\left(\frac{\partial f}{\partial x} f(x) \right)_i = \sum_{j=1}^{n} \frac{\partial f_i}{\partial x_j} f_j(x).$$

  Therefore we get the claim thanks to (6.5) and (6.6) .

- Suppose that (6.3) describes a canonical Hamiltonian vector field on the canonical symplectic vector space $\mathbb{R}^{2n}$. Assume that the Hamiltonian $H \in \mathscr{F}(\mathbb{R}^{2n})$ is a cubic polynomial in the components of x, so that f is quadratic polynomial in the components of x. Then:

  1. The discretization (6.4) preserves the following perturbation of H:

  $$H(x,\varepsilon) := H(x) + \frac{\varepsilon}{3}\left\langle \, \mathsf{grad}_x H(x), \left(1 - \frac{\varepsilon}{2}\frac{\partial f}{\partial x}\right)^{-1} f(x) \right\rangle,$$

  that is, $H(\widetilde{x},\varepsilon) = H(x,\varepsilon)$. Here $\langle \cdot, \cdot \rangle$ is the standard Euclidean product.

  2. The discretization (6.4) preserves the measure form $R(x)\, \mathrm{d}x_1 \wedge \cdots \wedge \mathrm{d}x_{2n}$ where

  $$R(x) := \mathsf{det}\left(1 - \frac{\varepsilon}{2}\frac{\partial f}{\partial x}\right).$$

  As a consequence, the discretization (6.4) applied to a planar (i.e., $n = 1$) canonical Hamiltonian vector field with a cubic Hamiltonian provides a completely integrable two-dimensional map.

6.2.2 *The Kahan-Hirota-Kimura discretization of the Euler top*

► We illustrate the Kahan-Hirota-Kimura scheme for the three-dimensional Euler top, thus proving that the corresponding discrete system is completely integrable as well.

► The three-dimensional Euler top is governed by the following three quadratic ODEs:

$$\begin{cases} \dot{x}_1 = \alpha_1\, x_2\, x_3, \\ \dot{x}_2 = \alpha_2\, x_3\, x_1, \\ \dot{x}_3 = \alpha_3\, x_1\, x_2, \end{cases} \tag{6.7}$$

with parameters $\alpha_i \in \mathbb{R}$. System (6.7) is solvable (see Example 3.8), bi-Hamiltonian (see Example 2.4) and completely integrable (see Example 3.2) In particular, the Poisson structure is induced by the Lie algebra $\mathfrak{so}(3)$ and the following three functions are Poisson involutive integrals of motion:

$$F_1(x) := \alpha_2\, x_3^2 - \alpha_3\, x_2^2, \qquad F_2(x) := \alpha_3\, x_1^2 - \alpha_1\, x_3^2, \qquad F_3(x) := \alpha_1\, x_2^2 - \alpha_2\, x_1^2.$$

Clearly, only two of them are functionally independent because of $\alpha_1 F_1 + \alpha_2 F_2 + \alpha_3 F_3 = 0$.

► The Kahan-Hirota-Kimura discretization of (6.7) is:

$$\begin{cases} \widetilde{x}_1 - x_1 = \varepsilon\,\alpha_1(\widetilde{x}_2\,x_3 + x_2\,\widetilde{x}_3), \\ \widetilde{x}_2 - x_2 = \varepsilon\,\alpha_2(\widetilde{x}_3\,x_1 + x_3\,\widetilde{x}_1), \\ \widetilde{x}_3 - x_3 = \varepsilon\,\alpha_3(\widetilde{x}_1\,x_2 + x_1\,\widetilde{x}_2). \end{cases} \tag{6.8}$$

In this form it corresponds to the stepsize 2ε rather than ε. The above system gives the following explicit birational map, $x \mapsto \widetilde{x} := \varphi_\varepsilon(x)$,

$$\begin{cases} \widetilde{x}_1 = \dfrac{x_1 + 2\,\varepsilon\,\alpha_1\,x_2\,x_3 + \varepsilon^2\,x_1\,(-\alpha_2\,\alpha_3\,x_1^2 + \alpha_3\,\alpha_1\,x_2^2 + \alpha_1\,\alpha_2\,x_3^2)}{\Delta(x,\varepsilon)}, \\[2ex] \widetilde{x}_2 = \dfrac{x_2 + 2\,\varepsilon\,\alpha_2\,x_3\,x_1 + \varepsilon^2\,x_2\,(\alpha_2\,\alpha_3\,x_1^2 - \alpha_3\,\alpha_1\,x_2^2 + \alpha_1\,\alpha_2\,x_3^2)}{\Delta(x,\varepsilon)}, \\[2ex] \widetilde{x}_3 = \dfrac{x_3 + 2\,\varepsilon\,\alpha_3\,x_1\,x_2 + \varepsilon^2\,x_3\,(\alpha_2\,\alpha_3\,x_1^2 + \alpha_3\,\alpha_1 x_2^2 - \alpha_1\,\alpha_2\,x_3^2)}{\Delta(x,\varepsilon)}, \end{cases} \tag{6.9}$$

where

$$\Delta(x,\varepsilon) := 1 - \varepsilon^2\left(\alpha_2\,\alpha_3\,x_1^2 + \alpha_3\,\alpha_1\,x_2^2 + \alpha_1\,\alpha_2\,x_3^2\right) - 2\,\varepsilon^3\,\alpha_1\,\alpha_2\,\alpha_3\,x_1\,x_2\,x_3.$$

► The next claim establishes the complete integrability of the discrete system (6.8).

Theorem 6.1

1. *System (6.8) admits the integrals of motion given by the functions*

$$F_1(x,\varepsilon) := \frac{1 - \varepsilon^2\,\alpha_3\,\alpha_1 x_2^2}{1 - \varepsilon^2\,\alpha_1\,\alpha_2\,x_3^2},$$

$$F_2(x,\varepsilon) := \frac{1 - \varepsilon^2\,\alpha_1\,\alpha_2\,x_3^2}{1 - \varepsilon^2\,\alpha_2\,\alpha_3\,x_1^2},$$

$$F_3(x,\varepsilon) = \frac{1 - \varepsilon^2\,\alpha_2\,\alpha_3\,x_1^2}{1 - \varepsilon^2\,\alpha_3\,\alpha_1\,x_2^2}.$$

   There are only two independent integrals of motion since

$$F_1(x,\varepsilon)F_2(x,\varepsilon)F_3(x,\varepsilon) = 1.$$

2. *The map (6.9) is a Poisson map w.r.t. the following parametric family of Poisson brackets:*

$$\{x_i, x_j\} = c_i\,\varepsilon\,\alpha_j\,x_k\left(1 - \varepsilon^2\,\alpha_k\,\alpha_i\,x_j^2\right) - c_j\,\varepsilon\,\alpha_i\,x_k\left(1 - \varepsilon^2\,\alpha_j\,\alpha_k\,x_i^2\right), \tag{6.10}$$

where c_1, c_2, c_3 are arbitrary constants. Here (i,j,k) is any cyclic permutation of $(1,2,3)$.

Proof. We prove only the first claim. Any of the functions $F_i(x,\varepsilon)$, $i=1,2,3$, is an integral of motion if $F_i(x,\varepsilon)=F_i(\widetilde{x},\varepsilon)$, namely if

$$\left(1-\varepsilon^2\,\alpha_k\,\alpha_i\,\widetilde{x}_j^2\right)\left(1-\varepsilon^2\,\alpha_i\,\alpha_j\,x_k^2\right)=\left(1-\varepsilon^2\,\alpha_i\,\alpha_j\widetilde{x}_k^2\right)\left(1-\varepsilon^2\,\alpha_k\,\alpha_i x_j^2\right),$$

which is equivalent to

$$\alpha_j\left(\widetilde{x}_k^2-x_k^2\right)-\varepsilon\,\alpha_k\left(\widetilde{x}_j^2-x_j^2\right)=\varepsilon^2\,\alpha_i\alpha_j\alpha_k\left(\widetilde{x}_j^2\,x_k^2-x_j^2\,\widetilde{x}_k^2\right).$$

The above equation is

$$\alpha_j\,(\widetilde{x}_k+x_k)(\widetilde{x}_k-x_k)-\alpha_k\,(\widetilde{x}_j+x_j)(\widetilde{x}_j-x_j)=\varepsilon^2\,\alpha_i\alpha_j\alpha_k\,(\widetilde{x}_j x_k+x_j\widetilde{x}_k)\,(\widetilde{x}_k x_j-x_k\widetilde{x}_j)\,.$$

Using equations (6.8) on both sides of the latter formula, we arrive at

$$(\widetilde{x}_k+x_k)\,(\widetilde{x}_i x_j+x_i\widetilde{x}_j)-(\widetilde{x}_j+x_j)\,(\widetilde{x}_k x_i+x_k\widetilde{x}_i)=(\widetilde{x}_i-x_i)\,(\widetilde{x}_k x_j-x_k\widetilde{x}_j)\,,$$

which is an algebraic identity. ■

► Remarks:

- The relation between the functions $F_i(x,\varepsilon)$ and the continuous integrals of motion $F_i(x)$ is the following: $F_i(x,\varepsilon)=1+\varepsilon^2\,\alpha_i\,F_i(x)+O(\varepsilon^4)$.
- The Poisson brackets (6.10) provide a polynomial ε-deformation of the continuous bi-Hamiltonian structure induced by $\mathfrak{so}(3)$.

► To conclude we present two plots showing the discrete orbits obtained by discretizing the Euler top ODEs in two different ways.

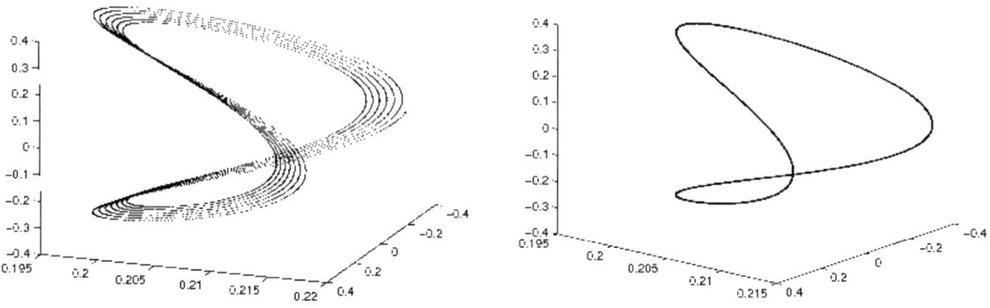

Fig. 6.1. Explicit Euler scheme (left) and Kahan-Hirota-Kimura scheme (right) applied to the Euler top ODEs.

On the left we have used the explicit Euler method and we see a spiraling trajectory, a typical non-integrable behavior. On the right we have used the Kahan-Hirota-Kimura discretization and we see that the orbit lies on a closed curve. Such a curve is indeed an elliptic curve given by the intersection of two quadrics.

6.2.3 Explicit solutions

► In Example 3.8 we derived explicit solutions of the continuous Euler top. We now construct the explicit solutions of the discrete integrable system (6.8), thus showing its solvability.

- We consider only the discretization of continuous solutions corresponding to the first Ansatz (see Example 3.8, formula (3.18)), namely

$$x_1 = a\,\frac{\vartheta_2(\nu\, t)}{\vartheta_4(\nu\, t)}, \qquad x_2 = b\,\frac{\vartheta_1(\nu\, t)}{\vartheta_4(\nu\, t)}, \qquad x_3 = c\,\frac{\vartheta_3(\nu\, t)}{\vartheta_4(\nu\, t)}. \tag{6.11}$$

  This is the general solution of the continuous three-dimensional Euler top for those initial data which correspond to $F_1, F_2 > 0$ and $F_3 < 0$, provided that

$$a^4 = \frac{F_2\, F_3}{\alpha_2\alpha_3}, \qquad b^4 = -\frac{F_1\, F_3}{\alpha_1\alpha_3}, \qquad c^4 = -\frac{F_1\, F_2}{\alpha_1\alpha_2}.$$

  Here F_1, F_2, F_3 are the integrals of motion. Similar arguments will apply for the second Ansatz (3.23).

- In the discrete case we consider again an Ansatz given by

$$x_1 = a\,\frac{\vartheta_2(\nu\, t)}{\vartheta_4(\nu\, t)}, \qquad x_2 = b\,\frac{\vartheta_1(\nu\, t)}{\vartheta_4(\nu\, t)}, \qquad x_3 = c\,\frac{\vartheta_3(\nu\, t)}{\vartheta_4(\nu\, t)}, \tag{6.12}$$

  which corresponds to the case $F_2 > 1$. Here F_2 is the discrete integral of motion given in Theorem 6.1.

- According to (6.12) we observe immediately that the discrete equations of motion (6.8) have a form which is similar to addition formulas (3.15-3.17):

$$\frac{\vartheta_1(u+v)}{\vartheta_4(u+v)} - \frac{\vartheta_1(u-v)}{\vartheta_4(u-v)} = \left(\frac{\vartheta_2(u+v)}{\vartheta_4(u+v)}\frac{\vartheta_3(u-v)}{\vartheta_4(u-v)} + \frac{\vartheta_2(u-v)}{\vartheta_4(u-v)}\frac{\vartheta_3(u+v)}{\vartheta_4(u+v)}\right)\frac{\vartheta_1(v)\vartheta_4(v)}{\vartheta_2(v)\vartheta_3(v)}, \tag{6.13}$$

$$\frac{\vartheta_2(u+v)}{\vartheta_4(u+v)} - \frac{\vartheta_2(u-v)}{\vartheta_4(u-v)} = -\left(\frac{\vartheta_1(u+v)}{\vartheta_4(u+v)}\frac{\vartheta_3(u-v)}{\vartheta_4(u-v)} + \frac{\vartheta_1(u-v)}{\vartheta_4(u-v)}\frac{\vartheta_3(u+v)}{\vartheta_4(u+v)}\right)\frac{\vartheta_1(v)\vartheta_3(v)}{\vartheta_2(v)\vartheta_4(v)}, \tag{6.14}$$

$$\frac{\vartheta_3(u+v)}{\vartheta_4(u+v)} - \frac{\vartheta_3(u-v)}{\vartheta_4(u-v)} = -\left(\frac{\vartheta_1(u+v)}{\vartheta_4(u+v)}\frac{\vartheta_2(u-v)}{\vartheta_4(u-v)} + \frac{\vartheta_1(u-v)}{\vartheta_4(u-v)}\frac{\vartheta_2(u+v)}{\vartheta_4(u+v)}\right)\frac{\vartheta_1(v)\vartheta_2(v)}{\vartheta_3(v)\vartheta_4(v)}. \tag{6.15}$$

We set the values of arguments of theta functions in formulas (6.13-6.15) as $u - v = \nu\, t$, $u + v = \nu(t + 2\varepsilon)$.

► We now present a complete construction of solutions.

- Formulas (6.12) substituted into (6.8) give

$$\begin{cases} \widetilde{x}_1 - x_1 = -\dfrac{a}{b\,c}\dfrac{\vartheta_1(\nu\,\varepsilon)\vartheta_3(\nu\,\varepsilon)}{\vartheta_2(\nu\,\varepsilon)\vartheta_4(\nu\,\varepsilon)}\,(\widetilde{x}_2\,x_3 + x_2\,\widetilde{x}_3), \\[2ex] \widetilde{x}_2 - x_2 = \dfrac{b}{c\,a}\dfrac{\vartheta_1(\nu\,\varepsilon)\vartheta_4(\nu\,\varepsilon)}{\vartheta_2(\nu\,\varepsilon)\vartheta_3(\nu\,\varepsilon)}\,(\widetilde{x}_3\,x_1 + x_3\,\widetilde{x}_1), \\[2ex] \widetilde{x}_3 - x_3 = -\dfrac{c}{a\,b}\dfrac{\vartheta_1(\nu\,\varepsilon)\vartheta_2(\nu\,\varepsilon)}{\vartheta_3(\nu\,\varepsilon)\vartheta_4(\nu\,\varepsilon)}\,(\widetilde{x}_1\,x_2 + x_1\,\widetilde{x}_2). \end{cases}$$

- Therefore (6.12) gives a solution if

$$\varepsilon\,\alpha_1 = -\frac{a}{b\,c}\frac{\vartheta_1(\nu\,\varepsilon)\vartheta_3(\nu\,\varepsilon)}{\vartheta_2(\nu\,\varepsilon)\vartheta_4(\nu\,\varepsilon)},$$

$$\varepsilon\,\alpha_2 = \frac{b}{c\,a}\frac{\vartheta_1(\nu\,\varepsilon)\vartheta_4(\nu\,\varepsilon)}{\vartheta_2(\nu\,\varepsilon)\vartheta_3(\nu\,\varepsilon)},$$

$$\varepsilon\,\alpha_3 = -\frac{c}{a\,b}\frac{\vartheta_1(\nu\,\varepsilon)\vartheta_2(\nu\,\varepsilon)}{\vartheta_3(\nu\,\varepsilon)\vartheta_4(\nu\,\varepsilon)}.$$

These are three equations for five unknowns (a, b, c, ν, q). As in the continuous time case, we use them to express the amplitudes a, b, c through ν, q:

$$a^2 = -\frac{1}{\varepsilon^2\alpha_2\,\alpha_3}\frac{\vartheta_1^2(\nu\,\varepsilon)}{\vartheta_3^2(\nu\,\varepsilon)}, \qquad b^2 = \frac{1}{\varepsilon^2\alpha_1\,\alpha_3}\frac{\vartheta_1^2(\nu\,\varepsilon)}{\vartheta_4^2(\nu\,\varepsilon)}, \qquad c^2 = -\frac{1}{\varepsilon^2\alpha_1\,\alpha_2}\frac{\vartheta_1^2(\nu\,\varepsilon)}{\vartheta_2^2(\nu\,\varepsilon)}.$$

- The missing two relations come from the integrals of motion. We proceed as in the continuous case and derive the formulas

$$\begin{aligned} 1 &= \frac{\vartheta_1^2(\nu t)}{\vartheta_4^2(\nu t)}\frac{\vartheta_3^2}{\vartheta_2^2} + \frac{\vartheta_2^2(\nu t)}{\vartheta_4^2(\nu t)}\frac{\vartheta_4^2}{\vartheta_2^2} = \frac{x_2^2}{b^2}\frac{\vartheta_3^2}{\vartheta_2^2} + \frac{x_1^2}{a^2}\frac{\vartheta_4^2}{\vartheta_2^2} \\ &= x_2^2\,\varepsilon^2\alpha_1\,\alpha_3\,\frac{\vartheta_4^2(\nu\,\varepsilon)\vartheta_3^2}{\vartheta_1^2(\nu\,\varepsilon)\vartheta_2^2} - x_1^2\,\varepsilon^2\alpha_2\,\alpha_3\,\frac{\vartheta_3^2(\nu\,\varepsilon)\vartheta_4^2}{\vartheta_1^2(\nu\,\varepsilon)\vartheta_2^2}, \end{aligned} \tag{6.16}$$

$$\begin{aligned} 1 &= \frac{\vartheta_1^2(\nu t)}{\vartheta_4^2(\nu t)}\frac{\vartheta_2^2}{\vartheta_3^2} + \frac{\vartheta_3^2(\nu t)}{\vartheta_4^2(\nu t)}\frac{\vartheta_4^2}{\vartheta_3^2} = \frac{x_2^2}{b^2}\frac{\vartheta_2^2}{\vartheta_3^2} + \frac{x_3^2}{c^2}\frac{\vartheta_4^2}{\vartheta_3^2} \\ &= x_2^2\,\varepsilon^2\alpha_1\,\alpha_3\,\frac{\vartheta_4^2(\nu\,\varepsilon)\vartheta_2^2}{\vartheta_1^2(\nu\,\varepsilon)\vartheta_3^2} - x_3^2\,\varepsilon^2\alpha_1\,\alpha_2\,\frac{\vartheta_2^2(\nu\,\varepsilon)\vartheta_4^2}{\vartheta_1^2(\nu\,\varepsilon)\vartheta_3^2}, \end{aligned} \tag{6.17}$$

and the linearly dependent one:

$$\begin{aligned} 1 &= \frac{\vartheta_3^2(\nu t)}{\vartheta_4^2(\nu t)}\frac{\vartheta_3^2}{\vartheta_4^2} - \frac{\vartheta_2^2(\nu t)}{\vartheta_4^2(\nu t)}\frac{\vartheta_2^2}{\vartheta_4^2} = \frac{x_3^2}{c^2}\frac{\vartheta_3^2}{\vartheta_4^2} - \frac{x_1^2}{a^2}\frac{\vartheta_2^2}{\vartheta_4^2} \\ &= -x_3^2\,\varepsilon^2\alpha_1\,\alpha_2\,\frac{\vartheta_2^2(\nu\,\varepsilon)\vartheta_3^2}{\vartheta_1^2(\nu\,\varepsilon)\vartheta_4^2} + x_1^2\,\varepsilon^2\alpha_2\,\alpha_3\,\frac{\vartheta_3^2(\nu\,\varepsilon)\vartheta_2^2}{\vartheta_1^2(\nu\,\varepsilon)\vartheta_4^2}. \end{aligned} \tag{6.18}$$

- Now we note that (6.16) can be written as $C_1\,x_1^2 + C_2\,x_2^2 = 1$ with coefficients

$$C_1 := -\varepsilon^2\alpha_2\,\alpha_3\,\frac{\vartheta_3^2(\nu\,\varepsilon)\vartheta_4^2}{\vartheta_1^2(\nu\,\varepsilon)\vartheta_2^2}, \qquad C_2 := \varepsilon^2\alpha_1\,\alpha_3\,\frac{\vartheta_4^2(\nu\,\varepsilon)\vartheta_3^2}{\vartheta_1^2(\nu\,\varepsilon)\vartheta_2^2},$$

and that these coefficients satisfy, due to the quadratic relation (3.10) for theta functions, the following linear relation with constant coefficients:

$$\alpha_1\,C_1 + \alpha_2\,C_2 = \varepsilon^2\alpha_1\,\alpha_2\,\alpha_3.$$

Thus, C_1, C_2 satisfy the following system:

$$\left\{\begin{array}{l} C_1\,x_1^2 + C_2\,x_2^2 = 1, \\ \alpha_1 C_1 + \alpha_2 C_2 = \varepsilon^2\alpha_1\,\alpha_2\,\alpha_3. \end{array}\right.$$

Its solution gives two (dependent) integrals of motion

$$C_1 = \frac{\alpha_2(1 - \varepsilon^2\alpha_1\,\alpha_3\,x_2^2)}{\alpha_2\,x_1^2 - \alpha_1\,x_2^2}, \qquad C_2 = \frac{\alpha_1(1 - \varepsilon^2\alpha_2\,\alpha_3\,x_1^2)}{\alpha_1\,x_2^2 - \alpha_2\,x_1^2}. \tag{6.19}$$

We can easily express both integrals (6.19) through the more symmetric integral F_3. We find:

$$1 - F_3 = \frac{\varepsilon^2\alpha_2\,\alpha_3}{C_1} = -\frac{\vartheta_1^2(\nu\,\varepsilon)\vartheta_2^2}{\vartheta_3^2(\nu\,\varepsilon)\vartheta_4^2}, \qquad 1 - F_3^{-1} = \frac{\varepsilon^2\alpha_1\,\alpha_3}{C_2} = \frac{\vartheta_1^2(\nu\,\varepsilon)\vartheta_2^2}{\vartheta_4^2(\nu\,\varepsilon)\vartheta_3^2}.$$

One has $C_1 > 0$, $C_2 > 0$, $F_3 > 1$.

- Analogously (6.17) can be written as $D_2\, x_2^2 + D_3\, x_3^2 = 1$ with coefficients

$$D_2 := \varepsilon^2 \alpha_1\, \alpha_3\, \frac{\vartheta_4^2(\nu\,\varepsilon)\vartheta_2^2}{\vartheta_1^2(\nu\,\varepsilon)\vartheta_3^2}, \qquad D_3 := -\varepsilon^2 \alpha_1\, \alpha_2\, \frac{\vartheta_2^2(\nu\,\varepsilon)\vartheta_4^2}{\vartheta_1^2(\nu\,\varepsilon)\vartheta_3^2}.$$

Again, these coefficients satisfy, due to the quadratic relation (3.9) for theta functions, the following linear relation with constant coefficients:

$$\alpha_2\, D_2 + \alpha_3\, D_3 = \varepsilon^2 \alpha_1 \alpha_2 \alpha_3.$$

Thus, D_2, D_3 satisfy the following system:

$$\begin{cases} D_2\, x_2^2 + D_3\, x_3^2 = 1, \\ \alpha_2\, D_2 + \alpha_3\, D_3 = \varepsilon^2 \alpha_1\, \alpha_2\, \alpha_3. \end{cases}$$

We find another two (dependent) integrals of motion

$$D_2 = \frac{\alpha_3(1 - \varepsilon^2 \alpha_1\, \alpha_2\, x_3^2)}{\alpha_3\, x_2^2 - \alpha_2\, x_3^2}, \qquad D_3 = \frac{\alpha_2(1 - \varepsilon^2 \alpha_3 \alpha_1 x_2^2)}{\alpha_2\, x_3^2 - \alpha_3\, x_2^2}. \tag{6.20}$$

Both integrals (6.20) can be easily expressed through F_1. We find:

$$1 - F_1 = \frac{\varepsilon^2 \alpha_3\, \alpha_1}{D_2} = \frac{\vartheta_1^2(\nu\,\varepsilon)\vartheta_3^2}{\vartheta_4^2(\nu\,\varepsilon)\vartheta_2^2}, \qquad 1 - F_1^{-1} = \frac{\varepsilon^2 \alpha_1\, \alpha_2}{D_3} = -\frac{\vartheta_1^2(\nu\,\varepsilon)\vartheta_3^2}{\vartheta_2^2(\nu\,\varepsilon)\vartheta_4^2}.$$

One has $D_2 > 0$, $D_3 > 0$, $F_1 < 1$.

- Finally (6.18) is written as $E_1\, x_1^2 + E_3\, x_3^2 = 1$ with coefficients

$$E_1 := \varepsilon^2 \alpha_2\, \alpha_3\, \frac{\vartheta_3^2(\nu\,\varepsilon)\vartheta_2^2}{\vartheta_1^2(\nu\,\varepsilon)\vartheta_4^2}, \qquad E_3 := -\varepsilon^2 \alpha_1\, \alpha_2\, \frac{\vartheta_2^2(\nu\,\varepsilon)\vartheta_3^2}{\vartheta_1^2(\nu\,\varepsilon)\vartheta_4^2}.$$

These coefficients satisfy, due to the quadratic relation (3.8) for theta functions, the following linear relation with constant coefficients:

$$\alpha_1\, E_1 + \alpha_3\, E_3 = \varepsilon^2 \alpha_1\, \alpha_2\, \alpha_3.$$

Thus, E_1, E_3 satisfy the following system:

$$\begin{cases} E_1\, x_1^2 + E_3\, x_3^2 = 1, \\ \alpha_1\, E_1 + \alpha_3\, E_3 = \varepsilon^2 \alpha_1\, \alpha_2\, \alpha_3. \end{cases}$$

This leads to further two (dependent) integrals of motion

$$E_1 = \frac{\alpha_3(1 - \varepsilon^2 \alpha_1\, \alpha_2\, x_3^2)}{\alpha_3 x_1^2 - \alpha_1\, x_3^2}, \qquad E_3 = \frac{\alpha_1(1 - \varepsilon^2 \alpha_2\, \alpha_3\, x_1^2)}{\alpha_1\, x_3^2 - \alpha_3\, x_1^2}. \tag{6.21}$$

Both integrals (6.21) can be expressed through F_2. We find:

$$1-F_2=\frac{\varepsilon^2\alpha_1\,\alpha_2}{E_3}=-\frac{\vartheta_1^2(\nu\,\varepsilon)\vartheta_4^2}{\vartheta_2^2(\nu\,\varepsilon)\vartheta_3^2},\qquad 1-F_2^{-1}=\frac{\varepsilon^2\alpha_2\,\alpha_3}{E_1}=\frac{\vartheta_1^2(\nu\,\varepsilon)\vartheta_4^2}{\vartheta_3^2(\nu\,\varepsilon)\vartheta_2^2}.$$

One has $E_1<0$, $E_3>0$, $F_2>1$. The latter inequality $F_2>1$ characterizes the applicability of the Ansatz (6.12).

- The three integrals F_1, F_2, F_3 are obviously dependent, because $F_1F_2F_3=1$.
- We finish by expressing the quantities q, ν through the integrals of motion. We have:

$$k^2:=\frac{\vartheta_2^4}{\vartheta_3^4}=\frac{1-F_3^{-1}}{1-F_1},$$

which lies in $(0,1)$ because of $F_1<F_3^{-1}<1$, since $F_2>1$. Finally, the already obtained expression

$$\frac{\vartheta_1^2(\nu\,\varepsilon)\vartheta_3^2}{\vartheta_4^2(\nu\,\varepsilon)\vartheta_2^2}=1-F_1$$

is the desired expression of ν through integrals.

6.3 An integrable discretization of the rational $\mathfrak{su}(2)$ Gaudin system

► In this Section we present an integrable discretization of the Hamiltonian system of ODEs governing the rational $\mathfrak{su}(2)$ Gaudin model (see 5.25):

$$\dot{\mathsf{X}}_i=\left[\omega_i\,\mathsf{P}+\sum_{j=1}^{N}\mathsf{X}_j\,,\,\mathsf{X}_i\right],\qquad i=1,\ldots,N,\tag{6.22}$$

where ω_i, $i=1,\ldots,N$, are distinct complex or real parameters. Recall that:

- We denote by $\mathsf{X}_i:=\left(x_i^1,x_i^2,x_i^3\right)^{\top}$, $i=1,\ldots,N$, the coordinate functions (in the basis E_α of $\mathfrak{su}(2)$) on the i-th copy of $\mathfrak{su}(2)^*$. The global phase space is $\mathfrak{G}^*:=\bigoplus^N\mathfrak{su}(2)^*\simeq\mathbb{R}^{3N}$. Any X_i can be expressed as

$$\mathsf{X}_i=\sum_{\alpha=1}^{3}x_i^\alpha\,\mathsf{E}_\alpha,\qquad i=1,\ldots,N.$$

- P in (6.22) is a constant vector of $\mathbb{R}^3$ (or, equivalently, a constant $\mathfrak{su}(2)$ matrix).
- System (6.22) is Hamiltonian w.r.t. the Lie-Poisson structure on $\mathfrak{G}^*$,

$$\left\{x_i^\alpha,x_j^\beta\right\}=-\delta_{ij}\,\varepsilon_{\alpha\beta\gamma}\,x_i^\gamma,\qquad i,j=1,\ldots,N,\tag{6.23}$$

with Hamiltonian

$$H := \frac{1}{2}\sum_{\substack{i,j=1\\ i\neq j}}^{N} \langle\, \mathsf{X}_i, \mathsf{X}_j \,\rangle + \sum_{i=1}^{N} \omega_i \,\langle\, \mathsf{P}, \mathsf{X}_i \,\rangle \,.$$

- The complete integrability of the system is ensured thanks to the existence of $2N$ independent and Poisson involutive integrals of motion given by

$$C_k := \frac{1}{2}\langle\, \mathsf{X}_k, \mathsf{X}_k \,\rangle\,, \qquad H_k := \langle\, \mathsf{P}, \mathsf{X}_k \,\rangle + \sum_{\substack{j=1\\ j\neq i}}^{N} \frac{\langle\, \mathsf{X}_k, \mathsf{X}_j \,\rangle}{\omega_k - \omega_j}\,, \qquad k = 1, \ldots, N.$$

► In the next statement we present a completely integrable map discretizing the Hamiltonian flow (6.22). This is an example of an *ad-hoc* discretization scheme, namely a scheme which does not belong to any prescribed discretization procedure.

Theorem 6.2

The map defined by

$$\widetilde{\mathsf{X}}_i = (1 + \varepsilon\,\omega_i\,\mathsf{P})\left(1 + \varepsilon\sum_{j=1}^{N}\mathsf{X}_j\right)\mathsf{X}_i\left(1 + \varepsilon\sum_{j=1}^{N}\mathsf{X}_j\right)^{-1}(1 + \varepsilon\,\omega_i\,\mathsf{P})^{-1}\,, \tag{6.24}$$

with $i = 1, \ldots, N$*, is a Poisson map w.r.t. (6.23) and admits* $2N$ *independent and Poisson involutive integrals of motion given by the functions* C_k*,* $k = 1, \ldots, N$*, and*

$$H_k(\varepsilon) = \langle\, \mathsf{P}, \mathsf{X}_k \,\rangle + \sum_{\substack{j=1\\ j\neq k}}^{N} \frac{\langle\, \mathsf{X}_k, \mathsf{X}_j \,\rangle}{\omega_k - \omega_j}\left(1 + \frac{\varepsilon^2}{4}\omega_k\,\omega_j\,\langle\, \mathsf{P}, \mathsf{P} \,\rangle\right) - \frac{\varepsilon}{2}\sum_{\substack{j=1\\ j\neq k}}^{N} \langle\, \mathsf{P}, [\mathsf{X}_k, \mathsf{X}_j] \,\rangle\,,$$

with $k = 1, \ldots, N$.

Proof: We proceed by steps.

- The map (6.24) is the composition of two non-commuting conjugations: $\varphi_\varepsilon = \varphi_\varepsilon^{(2)} \circ \varphi_\varepsilon^{(1)}$, where

$$\varphi_\varepsilon^{(1)}: \qquad \mathsf{X}_i \mapsto \widehat{\mathsf{X}}_i = \left(1 + \varepsilon\sum_{j=1}^{N}\mathsf{X}_j\right)\mathsf{X}_i\left(1 + \varepsilon\sum_{j=1}^{N}\mathsf{X}_j\right)^{-1}\,, \tag{6.25}$$

$$\varphi_\varepsilon^{(2)}: \qquad \widehat{\mathsf{X}}_i \mapsto \widetilde{\mathsf{X}}_i = (1 + \varepsilon\,\omega_i\,\mathsf{P})\,\widehat{\mathsf{X}}_i\,(1 + \varepsilon\,\omega_i\,\mathsf{P})^{-1}\,, \tag{6.26}$$

with $i = 1, \ldots, N$.

- The Poisson property of the map φ_ε is a consequence of the Poisson property of the maps $\varphi_\varepsilon^{(1)}$ and $\varphi_\varepsilon^{(2)}$. In fact, $\varphi_\varepsilon^{(1)}$ is a shift along a Hamiltonian flow on $\mathfrak{G}^*$ w.r.t. the Hamiltonian $\sum_{j,k=1}^{N} \langle \mathsf{X}_j, \mathsf{X}_k \rangle$. On the other hand $\varphi_\varepsilon^{(2)}$ is a shift along a Hamiltonian flow on $\mathfrak{G}^*$ w.r.t. the Hamiltonian $\sum_{k=1}^{N} \langle \mathsf{P}, \omega_k \mathsf{X}_k^* \rangle$. Therefore the composition $\varphi_\varepsilon^{(2)} \circ \varphi_\varepsilon^{(1)}$ is a Poisson map on $\mathfrak{G}^*$.
- We now prove that the functions $H_k(\varepsilon)$ are integrals of motion of the map (6.24). Their independence is clear, while their involution w.r.t. the brackets (6.23) can be proved by a direct computation.
- Notice that the maps (6.25) and (6.26) imply, respectively, the following relations (use (5.20)):

$$\left\langle \widehat{\mathsf{X}}_i, \widehat{\mathsf{X}}_j \right\rangle = \langle \mathsf{X}_i, \mathsf{X}_j \rangle, \qquad \widehat{\mathsf{X}}_i + \frac{\varepsilon}{2} \sum_{j=1}^{N} \left[\widehat{\mathsf{X}}_i, \mathsf{X}_j \right] = \mathsf{X}_i + \frac{\varepsilon}{2} \sum_{j=1}^{N} \left[\mathsf{X}_j, \mathsf{X}_i \right], \tag{6.27}$$

and

$$\left\langle \mathsf{P}, \widetilde{\mathsf{X}}_j \right\rangle = \left\langle \mathsf{P}, \widehat{\mathsf{X}}_j \right\rangle, \qquad \widetilde{\mathsf{X}}_i + \frac{\varepsilon}{2} \omega_i \left[\widetilde{\mathsf{X}}_i, \mathsf{P} \right] = \widehat{\mathsf{X}}_i + \frac{\varepsilon}{2} \omega_i \left[\mathsf{P}, \widehat{\mathsf{X}}_i \right], \tag{6.28}$$

with $i, j = 1, \ldots, N$.

- The preservation of the functions $H_k(\varepsilon)$ follows from the following computation:

$$\begin{aligned} \widetilde{H}_k(\varepsilon) &= \left\langle \mathsf{P}, \widetilde{\mathsf{X}}_k \right\rangle + \sum_{\substack{j=1 \\ j\neq k}}^{N} \frac{\left\langle \widetilde{\mathsf{X}}_k, \widetilde{\mathsf{X}}_j \right\rangle}{\omega_k - \omega_j} \left(1 + \frac{\varepsilon^2}{4} \omega_k \omega_j \langle \mathsf{P}, \mathsf{P} \rangle \right) \\ &\quad - \frac{\varepsilon}{2} \sum_{\substack{j=1 \\ j\neq k}}^{N} \left\langle \mathsf{P}, \left[\widetilde{\mathsf{X}}_k, \widetilde{\mathsf{X}}_j \right] \right\rangle \\ &= \left\langle \mathsf{P}, \widehat{\mathsf{X}}_k \right\rangle + \sum_{\substack{j=1 \\ j\neq k}}^{N} \frac{\left\langle \widehat{\mathsf{X}}_k, \widehat{\mathsf{X}}_j \right\rangle}{\omega_k - \omega_j} \left(1 + \frac{\varepsilon^2}{4} \omega_k \omega_j \langle \mathsf{P}, \mathsf{P} \rangle \right) \\ &\quad + \frac{\varepsilon}{2} \sum_{\substack{j=1 \\ j\neq k}}^{N} \left\langle \mathsf{P}, \left[\widehat{\mathsf{X}}_k, \widehat{\mathsf{X}}_j \right] \right\rangle \end{aligned}$$

where we used (6.28). Now, by using (6.27), we get $\widetilde{H}_k(\varepsilon) = H_k(\varepsilon)$ as desired.

The Theorem is proved. ■